Solar System Astronomy Activities Manual

Patrick Hall

York University

Kendall Hunt
publishing company

www.kendallhunt.com
Send all inquiries to:
4050 Westmark Drive
Dubuque, IA 52004-1840

ISBN: 978-1-5249-7898-3

Published in the United States of America

Contents

Properties of a Transiting Extrasolar Planet

NAME ______________________________ ID# ______________________________

DATE ______________________________ LAB SECTION# ____________________

Write your answers on each page of this lab and hand the entire lab in.

In this lab you will study a transiting extrasolar planet in a circular orbit around a main sequence star. You will use the star's temperature, luminosity, and motion and the lightcurve of one transit of the planet to learn as much as you can about the planet's physical parameters. Along the way, you will compare your planet to planets in our solar system.

Below we have illustrated what happens during a transit: a planet (small circle) moves between us and the star it orbits (big circle). As time goes on after a transit starts, the lightcurve of the star (bottom) shows that an increasing amount of the star's light is blocked until a maximum depth is reached (time t_2, Second Contact). The planet begins moving out from in front of the star at time t_3 (Third Contact), and the transit ends at time t_4.

1. Your lab coordinator will give you a unique chart of a simulated lightcurve of one transit of a unique planet and star. Near the bottom of the chart, the planet is given a number.

   Record the number of your planet here: __________

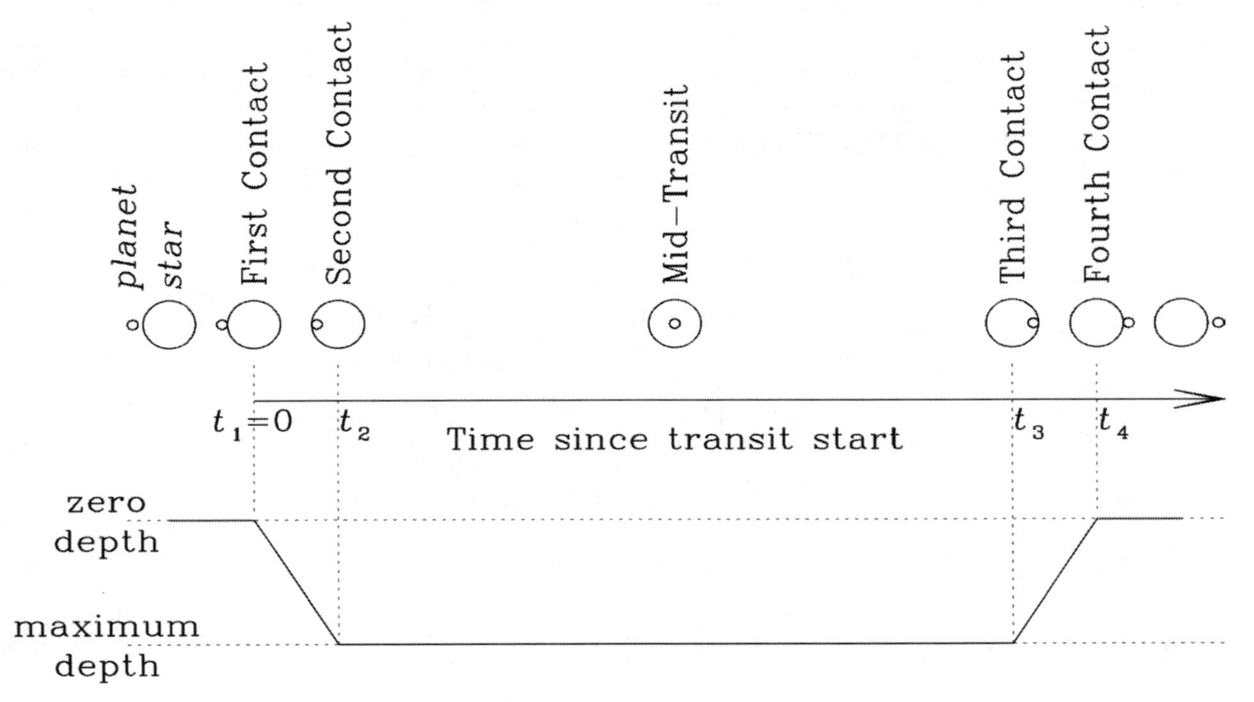

Figure 1: A planet (small circle) transiting across the exact center of its star (large circle).

Outline

Here is the information you will determine about your planet and star in this lab:

- Find the luminosity, mass, and radius of your star from its temperature.
- Find the radius of your planet from its maximum transit depth.
- Find the distance between your planet and star using Newton's version of Kepler's 3rd Law.
- Find what part of your star your planet transited from the time it took to transit.
- Find your planet's mass from your star's Doppler motion.
- Find your planet's gravity and density from its mass and radius.
- Find the cloud-top temperature on your planet (extra credit).
- Summarize how your planet compares to planets in our solar system.

From Temperature to Luminosity, Mass, and Radius of Your Star

Your chart includes the temperature of your star in Kelvin. Because your star is a main sequence star, its temperature is uniquely related to its radius and mass. Using Table 1, look up the luminosity LStar of your star relative to the Sun. (For example, a star with $T_{\text{Star}} = 10,000$ K has $L_{\text{Star}} = 56$, meaning that it emits 56 times more energy than the Sun.)

Table 1: Luminosities and Masses of Main Sequence Stars Relative to the Sun

Star's Temperature T_{Star}	4000 K	5000 K	5500 K	6000 K	8000 K	10000 K
Star's Luminosity L_{Star}	0.11	0.28	0.80	1.4	11	56
Star's Mass M_{Star}	0.54	0.75	0.90	1.1	2.0	2.9

We can use your star's luminosity and temperature to find its radius relative to the Sun's radius. We know that luminosity is proportional to area times temperature^4, that the area of a star is 4π times radius^2, and that the Sun has $L_{\text{Sun}} = 1$ and $T_{\text{Sun}} = 5780$ K. The ratio of your star's luminosity to the Sun's luminosity can therefore be used to get your star's radius relative to the Sun's:

$$\frac{L_{\text{Star}}}{L_{\text{Sun}}} = \frac{L_{\text{Star}}}{1} = \frac{4\pi(\text{Star's Radius})^2 T_{\text{Star}}^4}{4\pi(\text{Sun's Radius})^2 (5780\,\text{k})^4} \rightarrow \frac{\text{Star's Radius}}{\text{Sun's Radius}} = \sqrt{L_{\text{Star}}} \Big/ \left(\frac{T_{\text{Star}}}{5780\,\text{K}}\right)^2$$

2. Take the square root of L_{Star} and call it A.

$$A = \sqrt{L_{\text{Star}}} = \_\_\_\_\_\_\_ \qquad \text{(A)}$$

3. Divide T_{Star} by 5780, square the result, and call it B.

$$B = \left(\frac{T_{\text{Star}}}{5780}\right)^2 = \_\_\_\_\_\_\_ \qquad \text{(B)}$$

4. Now divide A by B to get R_{Star}, your Star's radius in terms of the Sun's radius:

$$R_{\text{Star}} = A/B = \_\_\_\_\_\_\_ \qquad \text{(C)}$$

Radius of Your Planet

The maximum depth of a planet's transit gives the planet's area as a fraction of the star's area. Because the area of a circle is $\pi(\text{Radius})^2$, the radius of the planet as a fraction of the star's radius is the square root of the depth:

$$\text{Depth} = \frac{\text{Planet Area}}{\text{Star Area}} = \frac{\pi(\text{Planet Radius})^2}{\pi(\text{Star Radius})^2} \rightarrow \frac{\text{Planet Radius}}{\text{Star Radius}} = \sqrt{\text{Depth}}$$

5. Measure the maximum depth of your planet's transit on the chart you were given. Note that the numbers increase DOWNWARDS on the graph. You should measure the depth to within at least 0.0001 (at least four numbers to the right of the decimal point, but no more than five).

   Maximum Depth = _._ _ _ _ _ (D)

6. Take the square root of the depth to find your Planet's radius relative to your Star's radius (call it E). Write down five numbers to the right of the decimal point.

   $\sqrt{\text{Maximum Depth}}$ = _._ _ _ _ _ (E)

7. It's easier to understand the planet's size relative to Earth's size. The Sun has a radius equal to 109 times the Earth's radius. Therefore we just have to multiply by 109 times the size of your Star relative to the Sun (C) to find the planet's radius relative to the Earth's radius (F):

   Planet's Radius relative to Earth's Radius $= 109 \times C \times E =$ _ _._ _ (F)

   (In your answer you should have only one or two numbers to the *left* of the decimal point and should keep two numbers to the right of the decimal point.)

8. Use Table 2 to answer the following question comparing your planet's radius to the radii of some planets in our Solar System.

   My planet's radius is in between that of the planets ________________ and ________________. (For example, if your planet's radius relative to Earth's was 0.7, the correct answer is "between Mercury and Venus", because those are the planets with radii closest to yours.)

Table 2: Some Useful Properties of Some Planets in our Solar System

Planet	Radius/ Earth's	Mass/ Earth's	Density in grams/cm^3	Semi-major Axis / Star's Radius	Gravity / Earth's	Cloud-top Temperature
Mercury	0.38	0.055	5.43	83.3	0.38	449 K (176 C)
Venus	0.95	0.815	5.25	155.6	0.90	329 K (56 C)
Earth	1.00	1	5.52	215.2	1	279 K (6 C)
Jupiter	11.19	317.9	1.33	1120.0	2.54	123 K (–150 C)
Saturn	9.46	95.18	0.70	2053.0	1.06	91 K (–183 C)
Neptune	3.81	17.13	1.64	6469.0	1.18	51 K (–222 C)

Distance of Your Planet from its Star

Newton's version of Kepler's Third Law states that for planets in circular orbits around a star of mass M_{Star}, the planet's orbital radius a and its orbital period P are related by

$$a^3 \propto M_{Star} \times P^2$$

as long as the mass of the planets is much less than M_{Star}. Specifically, if a is measured relative to our own Sun's radius and P is in Earth days, then

$$\left(\frac{a}{R_{Sun}}\right)^3 = 74.7 \times M_{Star} \times (P_{days})^2$$

9. Look at the bottom of your individual chart and find the number of days between each transit; that is the period P_{days} of your planet's orbit, measured in Earth days. Square P_{days}, multiply the result by 74.7 times M_{Star} (from Table 1), and call the result G.

$$G = 74.7 \times M_{Star} \times \left(P_{days}\right)^2 = \_\_\_\_\_\_\_\_\_\_ \qquad \text{(G)}$$

10. Take the cube root of G and divide it by the size of your star (C) to get a, your planet's orbital radius in units of the star's radius. (Taking a cube root is the same as raising to the power of 0.33.)

$$a = \frac{\sqrt[3]{G}}{C} = \_\_\_\_\_\_\_\_ = \_\_\_\_\_\_\_\_ \qquad \text{(H)}$$

11. Which solar system planet in Table 2 has an orbital radius in units of its star's radius which is most similar to your planet's?

What Part of Your Star Did Your Planet Cross?

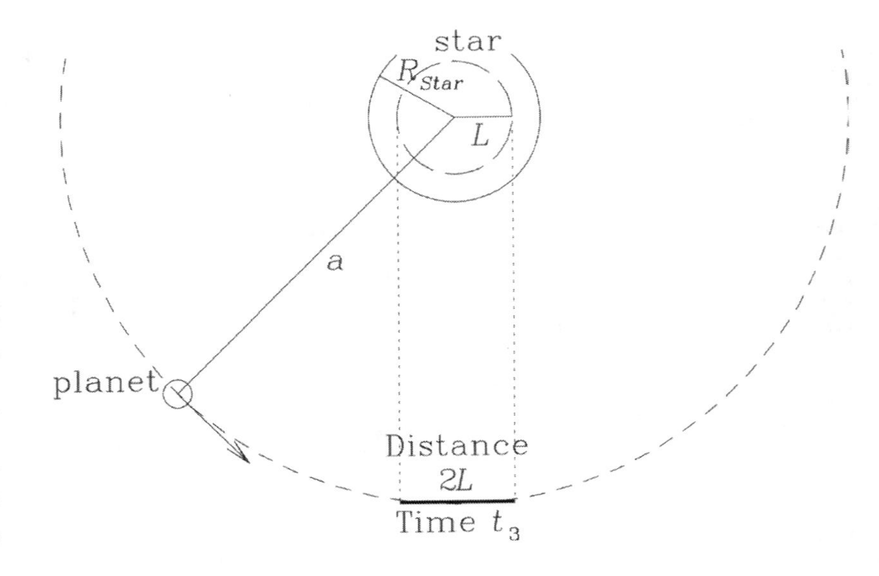

Figure 2: A planet orbiting its star, looking top-down onto the orbit. Earth is in the direction of the bottom of the page. A transit occurs when the planet is between the star and Earth. The star has diameter $2R_{Star}$ and the planet travels distance $2L$ during the transit. If the planet transits across the center of its star, $2L=2R_{Star}$, and the duration of the transit, t_3, is the maximum possible. If a planet transits off-center, $2L<2R_{Star}$, and the transit is shorter.

The distance around a circular orbit is 2π times the radius of the orbit. So your planet travels a distance $2\pi a$ in time P_{days} (one orbital period), which means its velocity is $v = 2\pi a / P_{days}$. Your planet also travels a distance $2L$ in time t_3 (and distance $2L + 2R_{Planet}$ in time t_4), which means its velocity is $v = 2L/t_3$. Those two expressions for the velocity must equal each other, which happens if:

$$\text{Fraction of star's diameter crossed during transit} = 2L = \frac{0.131 \times a \times t_3}{P_{days}}$$

where t_3 is measured in hours and a is measured in units of the star's radius.

12. Measure t_3, the time of Third Contact (end of maximum depth) on your transit chart (bottom panel of your handout), in hours. Multiply t_3 by 0.131 and by your planet's orbital radius in units of the star's radius:

$$J = t_3 \times 0.131 \times a = \_\_\_\_\_\_ \times 0.131 \times \_\_\_\_\_\_ = \_\_\_\_\_\_ \qquad \text{(J)}$$

13. Look at the bottom of your individual chart and find the number of days between each transit; that is the period P_{days} of your planet's orbit, measured in Earth days. Divide J by P_{days} to get $2L$, the fraction of your star's diameter crossed by your planet during one transit:

$$2L = \frac{J}{P_{days}} = \_\_\_\_\_\_ = \_\_\_\_\_\_ \qquad \text{(L)}$$

Figure 3 shows half of your star. The straight lines across it show paths taken by transiting planets. Each path crosses the fraction of the star's diameter indicated on the left-hand axis.

14. Draw a straight line on Figure 3 indicating the path of your transiting planet across the face of your star, using the left- and right-hand axis labels as a guide. For example, if you found a value of $2L$=0.990, you would draw a straight line between lines b and c, but closer to c.

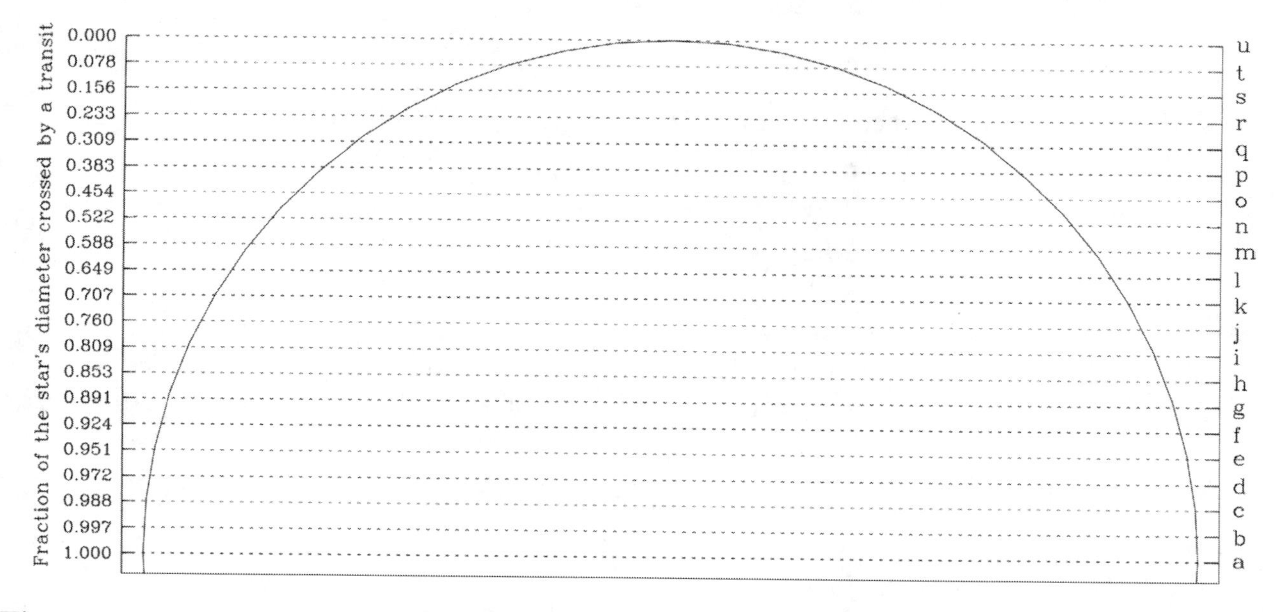

Figure 3: The lines show some of the paths planets can follow when transiting your star.

Mass of Your Planet

To measure the mass of your planet requires measuring the motion of the star around which it orbits. A planet in a circular orbit moves in a circle around the center of mass of the star- planet system, at the same time as the star moves in a much smaller circle around the same point. The planet and star are on exactly opposite sides of the center of mass at all times, which means that the star and planet velocities are related by:

$$v_p = v_s\, M_{\text{Star}} / M_p$$

We also know that the planet travels a distance 2πa in one orbital period P_y (measured in Earth years), so the planet's velocity also equals:

$$v_p = 2\pi a / P_y$$

Those two equations for v_p must equal each other, and if we combine them we get an expression for the planet's mass relative to the star's mass, which we will use below:

$$M_p / M_{\text{Star}} = v_s P_y / 2\pi a$$

The only complication is that we measure P in days, not years, which introduces another number we will combine with the factor of 2π in our calculations.

15. Measure v_s, the **maximum** radial velocity of the star in the top panel of your handout. Multiply that number by P_{days}, the period of your planet's orbit in Earth days (see Question 9). Then divide by H (your planet's orbital radius in units of the star's radius) times C (your star's radius in units of the Sun's radius). Call the result N.

$$N = \frac{v_s \times P_{\text{days}}}{H \times C} = \frac{\quad \times \quad}{\quad \times \quad} = \underline{\qquad\qquad} = \underline{\qquad\qquad} \qquad \text{(N)}$$

16. Divide N by 50,578 and call the result O; that is your planet's mass relative to its star.

$$O = \frac{N}{50{,}578} = \frac{\textit{Planet Mass}}{\textit{Star Mass}} = \_.\_\_\_\_\_\_ \qquad \text{(O)}$$

17. It is easier to understand a planet's mass if it is in Earth masses. The Sun has a mass of 332,746 Earth masses, so your star has a mass of (332, 746 × M_{Star}) Earth masses. You can multiply O by 332,746 × M_{Star} to get M, the mass of your planet relative to Earth's mass.

$$M = O \times 332{,}746 \times M_{\text{Star}} = \underline{\qquad} \times 332{,}746 \times \underline{\qquad} = \_\_\_.\_\_ \qquad \text{(M)}$$

18. Use Table 2 to answer the following question.

My planet's mass is in between that of the planets ______________ and ______________.

Density of Your Planet

Now that you know your planet's radius (size) and mass, you can determine its density (mass divided by volume) and compare it to high-density terrestrial planets and low-density giant planets in our solar system.

19. Cube your planet's radius relative to Earth (F) and call the result Q:

$$F^3 = F \times F \times F = Q = \_\_\_\_\_\_\_\_ \qquad \text{(Q)}$$

20. Multiply your planet's mass M by 5.5 and divide it by Q to get your planet's density R in grams per cubic centimeter (cm^3):

$$\frac{5.5 \times M}{Q} = R = \_\_\_\_\_\_ \qquad \text{(R)}$$

21. According to Table 2, your planet's density R is most similar to the density of (circle one):
    - A terrestrial planet like Mercury, Venus, or Earth
    - A gas giant planet like Jupiter
    - A low-density gas giant planet like Saturn
    - An ice giant planet like Neptune

How Much More Would You Weigh on Your Planet?

What you feel as weight is the force of Earth's gravity pulling on your body. The more mass you have, the more Earth pulls on you. The force of any planet's gravity per unit mass (one g) is proportional to the planet's mass divided by its radius squared: $g \propto M/R^2$. If you were to stand on another planet (or on a blimp in its outer atmosphere, if it doesn't have a solid surface), you would feel a different weight. Let's see how much more you would weigh on your planet.

22. Square your planet's radius relative to Earth (F) and call the result S:

$$F^2 = F \times F = S = \_\_\_\_\_\_\_\_ \qquad \text{(S)}$$

23. Divide your planet's mass M by S; the result—call it W—is your planet's surface gravity relative to Earth's:

$$\frac{M}{S} = W = \_\_\_\_\_\_\_\_ \qquad \text{(W)}$$

In other words, on your planet you would weigh W times as much as on Earth.

24. Use Table 2 to compare your planet's surface gravity to that of planets in our solar system:

    The surface gravity of my planet is greater than that of __________ but less than that of __________. (If there is no planet in Table 2 with a larger or smaller surface gravity than your planet's, put 'no planet in Table 2' in the appropriate blank.)

Summary of Your Planet

25. In one sentence, give your planet's properties as compared to planets in our solar system. For example: "My planet orbits at the distance of Venus but is bigger than Neptune, with a mass close to Jupiter's, giving it a terrestrial planet's density and a surface gravity four times Earth's."

____

____

____

Extra Credit

Cloud-Top Temperature of Your Planet

The cloud-top temperature of a fast-rotating planet is determined by the balance between how much heat the planet absorbs from the star and how much heat it emits into space. The size of the planet does not matter: a bigger planet absorbs more heat but also emits more heat. What does matter is how close the planet is to the Sun, and how much heat from the Sun it reflects rather than absorbs.

If we assume that the planet absorbs all the heat that hits it, we can estimate the temperature at the top of its atmosphere. (If the planet has no atmosphere, the cloud-top temperature will just be the surface temperature. But if the planet does have an atmosphere, the greenhouse effect can increase the surface temperature above the cloud-top temperature.)

The cloud-top temperature of a fast-rotating planet orbiting a star of temperature T_{Star} is

$$\text{Cloud-top Temperature} = \frac{0.7071 \times T_{\text{Star}}}{\sqrt{\text{Planet's orbital radius a, relative to the star's radius}}}$$

26. **[Extra Credit]** Calculate your planet's cloud-top temperature in Kelvin:

$$T = \frac{0.7071 \times T_{\text{Star}}}{\sqrt{H}} = \_\_\_\_\_\_\_\_\_\_ \text{K} \qquad \text{(T)}$$

27. **[Extra Credit]** Convert your planet's cloud-top temperature to degrees Celsius:

$$U = T - 273.15 = \_\_\_\_\_\_\_\_\_\_ \text{C} \qquad \text{(U)}$$

28. **[Extra Credit]** Use Table 2 to compare your planet's cloud-top temperature to that of planets in our solar system. (If there is no planet in Table 2 with a lower or higher cloud-top temperature than your planet's, put 'no planet in Table 2' in the appropriate blank.)

The cloud-top temperature of my planet is greater than that of __________ but less than that of __________.

Unit 1.3 Temperature Scales

Objective

To become acquainted with the relations between the three most common temperature scales: Celsius, Fahrenheit and Kelvin

Introduction

Temperature is a measure of the energy of the particles which make up a body or an environment, like a cup of water or the atmosphere of a star, or rock on the surface of a planet. The temperature is related to the average speed of atoms or molecules which make up an object. A hot object, like the water molecules in a cloud of steam, are moving rapidly and chaotically, with each molecule containing relatively large amounts of energy. By contrast, the low-velocity atoms of water ice are held stationary in a solid crystal structure. As the ice is heated, the velocity of the molecules begins to break apart those bonds, reducing the crystal structure to a liquid puddle and eventually an expanding gas cloud.

Three temperature scales are most commonly used in everyday life, science, and industry.

The **degree Celsius (°C)** temperature scale was devised by the Swedish astronomer Anders Celsius in 1742. This scale is based on the behavior of pure water, which freezes at 0 °C and boils at 100 °C under standard atmospheric conditions (at sea level on the Earth). Therefore, there are 100 degrees between these points. This scale is used throughout the science and almost in all countries of the world except the United States.

The **Kelvin (K)** temperature scale, named after the British Lord Kelvin (William Thomson, Baron Kelvin of Largs) is an extension of the degree Celsius scale down to *absolute zero or 0 K,* a hypothetical temperature at which all atomic and molecular motions cease. *Absolute zero* is the lowest temperature hypothetically possible at which no heat exists. On the Kelvin temperature scale, water freezes at 273 K (more precise value: 273.16 K) and boils at 373 K. Absolute zero (0 K) or –273 °C is the starting point for the Kelvin scale. Since nothing can be colder than 0 K, there are no negative temperatures on the Kelvin scale.

The step size of the Kelvin and the Celsius temperature scales is same, because water must be heated by 100 K or 100 °C to go from its freezing to melting point. Scientists throughout the world (including the United States) prefer the Kelvin scale because it is closely related to the physical meaning of the temperature.

Note: Temperatures on this scale are called "kelvin," *not* degrees kelvin. Further, kelvin is *not* capitalized, and the symbol (capital K) stands alone with no degree symbol.

The **degree Fahrenheit (°F)** temperature scale, now antiquated, is still used by many in the United States. The German physicist Gabriel Fahrenheit introduced this scale in the early 1700s and he intended 0 °F to represent the coldest temperature achievable at that time and 100 °F to represent the temperature of a healthy human body. As you might know, normal body temperature is closer to 98.6 °F, suggesting that when he conducted his experiment, either he was having a fever, or his thermometer was inaccurate. Lastly, it is believed that he might have used a cow's temperature instead of his own. On this scale, water freezes at 32 °F and boils at 212 °F. Therefore, there are 180 degrees between these points. The step size of the Fahrenheit degree is smaller than that of the Celsius degree (or 1 kelvin). In other words, a degree Celsius (or a kelvin) is 180/100 (which is 9/5 or 1.8 times) the size of the degree Fahrenheit. An increase of 1 kelvin is equivalent to a Fahrenheit temperature increase of nearly 2 degrees. Notice that the degree Fahrenheit is a non-metric temperature scale, while the degree Celsius and the Kelvin temperature scales are metric scales (based on multiples of 10).

Note that the United States is the only country that uses Fahrenheit temperatures for shelter-level (surface) weather observations. However, since July 1996 all surface temperature observations in the National Weather Service METAR/TAF reports are transmitted in degrees Celsius.

Equations and Constants

Fahrenheit to Celsius Conversion $$C = \frac{(F-32)}{1.8}$$

Fahrenheit to Kelvin Conversion $$K = \frac{(F-32)}{1.8} + 273.16$$

Celsius to Fahrenheit Conversion $$F = (C \times 1.8) + 32$$

Celsius to Kelvin Conversion $$K = C + 273.16$$

Kelvin to Fahrenheit Conversion $$F = (K - 273.16) \times 1.8 + 32$$

Kelvin to Celsius Conversion $$C = K - 273.16$$

NAME ______________________________ ID ______________________________

DUE DATE __________ LAB INSTRUCTOR __________ SECTION __________

Worksheet # 1

Answer the following questions related to the temperature scale

1. Which temperature scale or scales begin at zero?

2. Which temperature scale or scales allow for negative temperatures?

3. At what temperature does water freeze:

   On the Fahrenheit scale:

   On the Celsius scale:

   On the Kelvin scale:

4. A 1 degree temperature change on the Fahrenheit scale is equal to how many degree of temperature change on the Celsius scale?

5. Normal human body temperature is 98.6 °F; what is the healthy human body temperature:

   On the Celsius scale:

   On the Kelvin scale:

6. The color of a hot metal is directly related to the temperature of the metal. The coil on a stovetop burner will begin to glow a dim, deep red at 390 °C. At what temperature does a stove coil begin to glow:

   On the Celsius scale:

   On the Kelvin scale:

Continue....

7. The boiling point of oxygen – where oxygen transitions from a liquid puddle into a gas cloud – occurs at –183 °C. At what temperature does that process happen:

   On the Fahrenheit scale:

   On the Kelvin scale:

8. When temperatures drop below 63.15 K, nitrogen freezes, turning from liquid nitrogen into solid ice. At temperatures above 77.36 K, liquid nitrogen turns into a gas. On the dwarf planet Pluto, the wintertime temperature is low enough to freeze the nitrogen atmosphere, turning the atmosphere into a gentle snow of ice crystals. In the summertime, the temperature is high enough to turn nitrogen into a gas. If a thermometer on the surface of Pluto reads –214.15 °C, show whether Pluto's nitrogen will be frozen into ice, existing in puddles, or in a gaseous form making up an atmosphere.

9. The Sun's surface temperature is 5770 K. What is the temperature of the Sun:

   In Fahrenheit:

   In Celsius:

10. Absolute 0 on the Kelvin scale is 0 K. What is this in Fahrenheit?

11. At the beginning of the week, the temperature is measured at a chilly 40 °F. At the end of the week, the temperature has risen to 80 °F. Has the temperature doubled? Explain why of why not. (As a hint, consider look at question 10 and consider where the Fahrenheit scale begins).

Unit 1.2 Measuring Angles

Objective

To become familiar with angles, their measurements, conversion, and use in astronomy

Introduction

In geometry, any two connected points form a line. Two connected lines create an angle between them. Before the advent of telescopes (which allowed for detailed study of objects too small to see with the naked eye) astronomy as a science was limited to drawing star charts and carefully measuring the positions of planets against the background stars. Even with telescopes, careful studies of positions and measured angles allowed astronomers to determine the sizes of the planets and the distances between orbits.

Angles are measured in the unit of degrees. One full circle, rotation, or revolution contains 360 degrees. In much the same way as an hour is composed of smaller units called minutes, 1 degree is composed of smaller units called *arcminutes*. One arcminute is an incredibly small angle. Two lines which deviate by 1 arcminute would appear so close to one another that they be mistaken for one line to your naked eye. In fact, only after drawing the lines nearly 300 inches (25 feet) long would you see the ends of the lines separated by 1 inch, making an extremely long sliver of a triangle with a 1 inch long side.

Just as 1 minute of time is further broken down into 60 individual seconds, 1 arcminute is composed of 60 equally spaced subunits called *arcseconds*. Two lines which make an angle of 1 arcsecond to each other would have to be drawn out incredibly long before their deviation was noticeable. At 3.25 miles long, the ends of the two lines would lie one inch apart.

Angular diameter and angular separation are two critical concepts in astronomy and they both utilize the measurement of angles. The angular separation describes how distant two objects *appear* from one another in the sky. From our vantage point on Earth, the sky appears flat and two dimensional. The distance we see between stars is an angular separation, not an actual three dimensional, linear distance. For example, the three stars of Orion's belt appear very close to one another, making a straight line in the sky. In actuality, those three stars are each hundreds of light years away from the Earth and even further apart from one another. Their relative closeness (small angular separation) is an optical illusion. Along the same vein, when planets "align" during a conjunction and appear extremely close to one another in the sky, they are actually tens of millions of miles apart.

Angular diameters describe what percentage of your vision is occupied by an object. Human vision spans about 180° across. Something incredibly small – like the head of a pin from 100 yards away – takes up so little of your vision as to be invisible. The head of a pin viewed from a football field away would cover about 1 arcsecond of vision. The Great Wall of China – seen from orbit – is only about 20 arcminutes wide, still too small for the receptors in the eye to recognize.

Your fist – held out at arm's length – will cover roughly 10 degrees of your vision. In this way, angular diameter describes how large something appears to be rather than its absolute size.

Procedure

Measuring the angular separation between two lines requires a protractor. The standard protractor is a semi-circle marked off with ticks running from 0° to 180°. To measure the angle made by two lines, place the center of the protractor on the intersection point of the two lines.

Adjust the protractor until one of the lines is pointing directly at the angle of 0°. The second line will point to the reading representing the angle between the lines. It may make measurements easier to use a ruler and extend the length of the lines (which make the angle) so that you do not have to estimate where the measured line points.

Likewise, to draw an angle, begin by drawing two points and connecting them with a straight line. This will serve as the baseline. Place the center of the protractor at one end of the line, making sure that the line points to 0°. Mark a third point on your protractor at the desired angular separation in degrees and use the straight edge of a ruler to connect those points. Smaller units are one arcmin (1') and one arcsec (1").

Equations and Constants

$$1^\circ = 60'$$

$$1^\circ = 3600''$$

$$1' = \frac{1}{60}^\circ$$

$$1'' = \frac{1}{60}' = \frac{1}{3600}^\circ$$

NAME ____________________ ID ____________________

DUE DATE __________ LAB INSTRUCTOR ______________ SECTION __________

Worksheet # 1

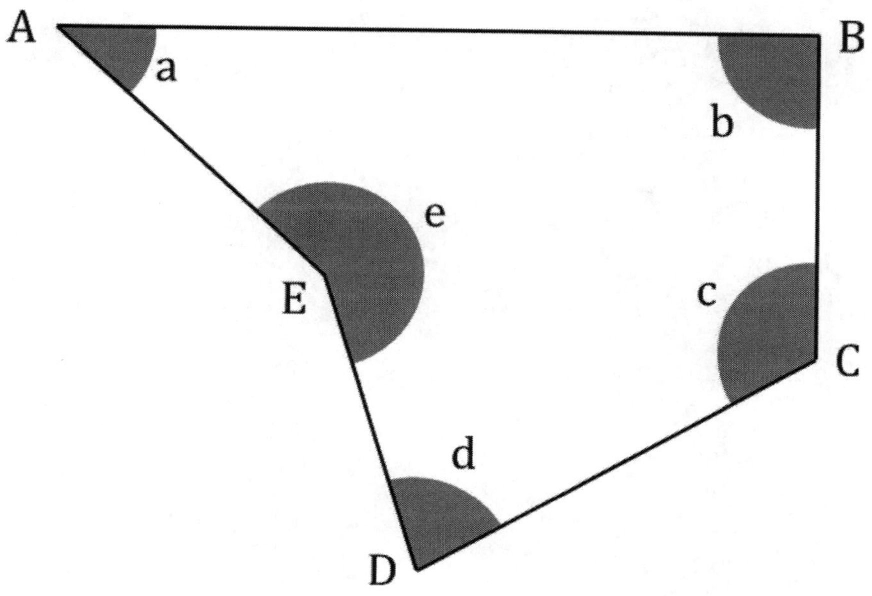

A B C D E a b c d e

Below is a 5 sided geometric shape made up of lines AB, BC, CD, DE, and EA. At the point where the lines meet they make angles a, b, c, d, and e, with the angle highlighted in gray. Use a protractor to measure the indicated angles and give an answer in degrees.

1. Angle **a** (the angle between lines AB and AE)

2. Angle **b** (the angle between lines AB and BC)

3. Angle **c**

4. Angle **d**

5. Angle **e**

NAME ______________________ ID ______________________

DUE DATE __________ LAB INSTRUCTOR ______________ SECTION __________

Worksheet # 2

Using a protractor, draw a pair of lines intersecting at the listed angle.

1. 17°

2. 68°

3. 91°

4. 166°

5. 345°

NAME ______________________ ID ______________________

DUE DATE __________ LAB INSTRUCTOR ______________ SECTION __________

Worksheet # 3

Use the conversions to determine the following angles.

1. In the Earth's sky, the moon has an angular diameter of 0.5°. What is the angular diameter of the Moon in arcminutes?

2. Through a telescope, two stars are separated by 2° 15' 35". What is their separation in arcseconds?

3. The stars Mizar and Alcor – both in the handle of the Big Dipper – are located 12' 21" from one another. What is that separation in arcseconds?

4. Arrange these angular diameters in order of largest to smallest angular diameter:

   a) 0.25°

   b) 90'

   c) 1000"

NAME ________________________________ ID ________________________________

DUE DATE __________ LAB INSTRUCTOR ______________ SECTION __________

Worksheet # 4

The four dashed circles below represent the orbits of the innermost planets, with S representing the position of the sun, V the position of Venus, E the position of Earth, and M the position of Mars at some given time.

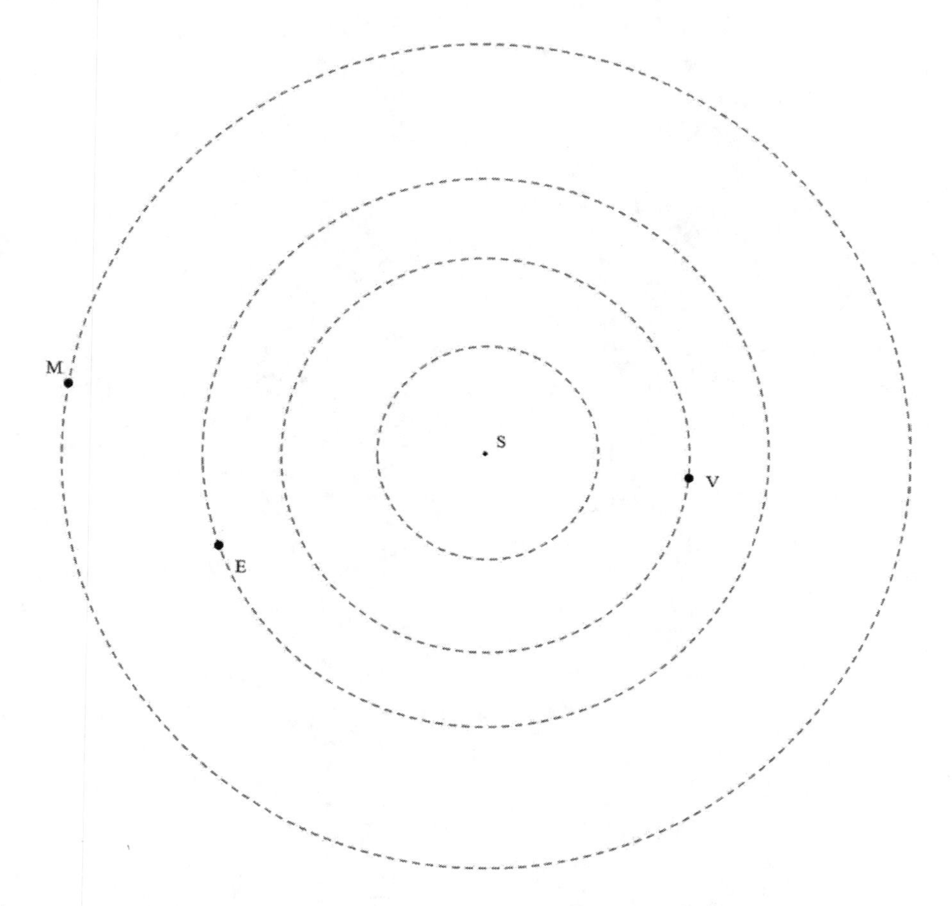

1. From the perspective of someone on Earth, how far apart would the Sun and Mars appear, in degrees? (What is the angle from M to E to S?) How far apart would the Sun and Venus appear, in degrees? (What is the angle from S to E to V?)

2. Someone on Earth determines that Mercury is 16° from the Sun. Mark Mercury's possible position in orbit with an ***M*** (Mercury's orbit is the innermost orbit.)

Motion on the Celestial Sphere

NAME ______________________ ID# ______________________

DATE ______________________

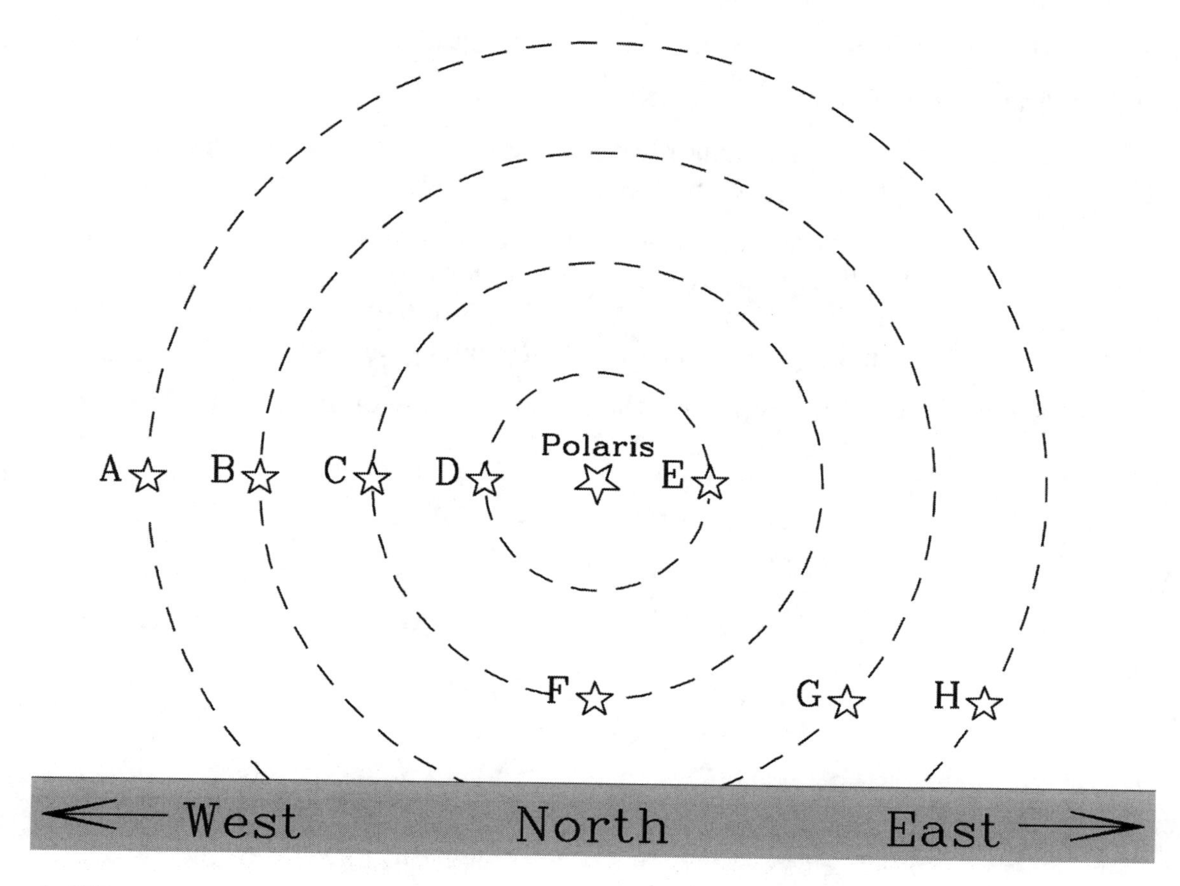

Figure 1: What someone in the northern hemisphere might see looking north at midnight at this time of year. The horizontal black line is your **horizon** (the grey area is part of the surface of the Earth). You can only see stars above your horizon.

Polaris (the North Star) is shown above, along with eight other stars. As the Earth rotates, all eight stars appear to move in circular paths around the North Celestial Pole (the NCP). The dashed lines show these circular paths.

1. Mark the location of the North Celestial Pole (NCP) in Figure 1.
2. Your **meridian** is the line on the celestial sphere going from due North through the NCP and continuing due South to stop at the point due South on your horizon. **Draw the part of your meridian visible** in Figure 1. Which star besides Polaris is on the meridian at the time shown? ______________________
3. The Earth rotates so that the Sun rises above the Earth's horizon in the East and sets below the Earth's horizon in the West. Therefore, in which direction do these stars appear to move

on their circular paths around the NCP? *(Clockwise|Counterclockwise)* Draw an arrow at each star showing the direction that star appears to move at the time shown in Figure 1. [Note: These stars move through 360° in a 24-hour period.]

4. List any stars that are moving straight down towards the horizon: ______________
   List any stars that are moving straight up away from the horizon: ______________
   List any stars that are moving horizontally (i.e., parallel to the horizon): ______________
5. Will all these stars be above the horizon for the same length of time each day? *(Yes|No)*
6. Among the stars A, B, C, and D, which one will set first? ______________
7. Stars that never set from a particular latitude are known as **circumpolar** stars. Which of the eight stars A through H are circumpolar? ______________
8. Figure 1 shows the sky at midnight. Six hours later (6 A.M.) the Sun will rise. Which of the stars A through H will be above the horizon at sunrise? ______________ At 6 P.M., six hours earlier than the time shown in Figure 1, the Sun had set and the sky had just gotten dark enough to see stars. Which of the stars A through H was visible then? ______________
9. Mark the meridian and the zenith for the tiny stick figure on the Earth in Figure 2.

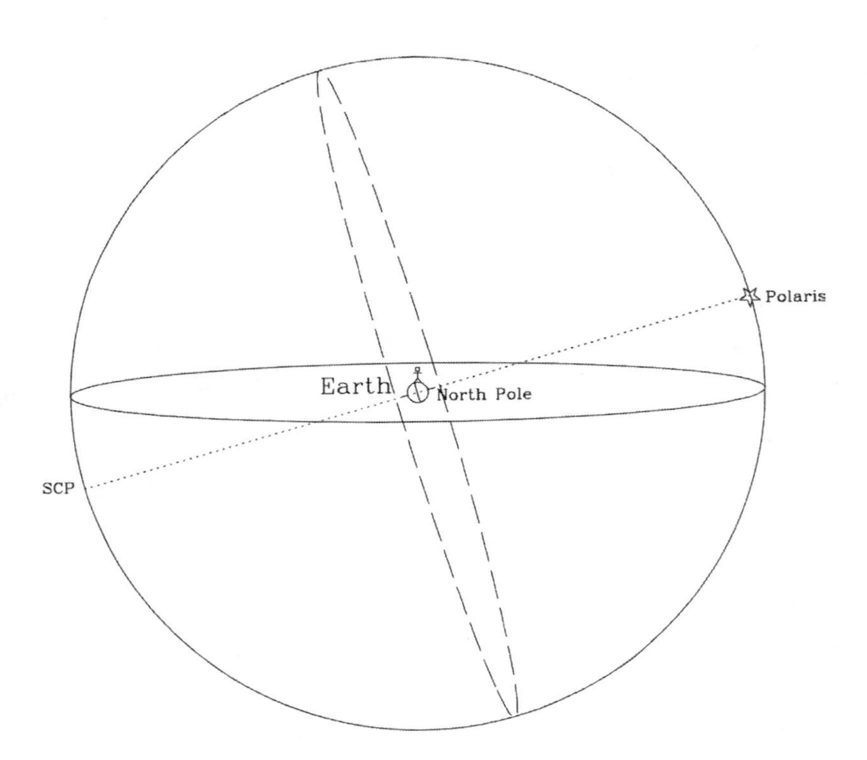

Figure 2: The same situation as shown in Figure 1, drawn in terms of our model of the **celestial sphere:** an imaginary spherical surface surrounding the rotating Earth. In reality, stars are located at many different distances from Earth, but we can imagine them as being projected onto the celestial sphere. Also, because the Earth spins on its axis (the line connecting Earth's North Pole and South Pole), the stars appear to move. However, relative to our point of view on Earth, the Earth doesn't move and we can instead imagine the celestial sphere rotating around the Earth on the same axis (the dotted line).

10. Mark the horizon for the stick figure in Figure 2. Indicate which parts of Figure 2 are above the horizon for that stick figure by shading only the parts that are below the horizon. (Hint: We are viewing the stick figure from slightly above its horizon.)

11. Draw on Figure 2 the circular path on the sky of any one of the stars in Figure 1, including arrows to indicate the direction of motion of the star on its path.

12. What is the dashed line girding the celestial sphere in Figure 2? ______________

13. **(Optional)** Shade in the region of the celestial sphere where stars that are *never visible* from the stick figure's location are found.

Motions of the Moon

NAME ______________________________ DATE ______________________________

Definitions

Define the following terms in your own words:

1. Eclipse

2. Conservation of Angular Momentum

3. Lunar Month (Synodic Month)

4. Sidereal Month

5. Roche Limit

6. Tidal Locking

NAME ______________________ DATE ______________________

Important Facts

1. Sketch the alignment of the Sun, Moon, and Earth for the following:
   a. a solar eclipse

   b. a lunar eclipse

   c. spring tides

   d. neap tides

2. a. The Moon's orbital period (sidereal month) is ______________days
   b. The Moon's phase period (lunar month) is ______________days
3. a. At what phase of the Moon does a solar eclipse occur? ______________
   b. At what phase of the Moon does a lunar eclipse occur? ______________
4. What is the current rate at which the Moon recedes from Earth under tidal influences? ______________ cm/yr
5. What are the only two worlds in the Solar System that are known to be tidally-locked to each other? ______________and______________

NAME ______________________________ DATE ______________________________

6. Complete the following table of lunar phases:

Phase	Sketch	Moonrise/Moonset	Orientation to Sun
New		**Moonrise:** **Moonset:**	
Waxing Crescent		**Moonrise:** **Moonset:**	
First Quarter		**Moonrise:** **Moonset:**	
Waxing Gibbous		**Moonrise:** Afternoon **Moonset:** After midnight	Between 90° to 180° east of the Sun
Full		**Moonrise:** **Moonset:**	
Waning Gibbous		**Moonrise:** **Moonset:**	
Last Quarter		**Moonrise:** **Moonset:**	
Waning Crescent		**Moonrise:** After midnight **Moonset:** Afternoon	< 90° west of the Sun

NAME ______________________________ DATE ______________________________

Critical Thinking Questions

1. Why is the lunar month different from the sidereal month? Draw a sketch to accompany your explanation.

2. Why are some solar eclipses total while others are annular? Were total eclipses or annular eclipses more common in the distant past? Which will be more common in the distant future? Explain your answers.

3. If the Moon is receding from Earth, then the angular momentum in its orbit is increasing. Where does it get this additional angular momentum?

Lunacy: Moon Phase Exercise

NAME ______________________________ DATE ______________________________

1. Complete the following moon phase chart with rising, setting and culmination times (sunset, sunrise, noon and midnight). Refer to the moon phase diagram on the preceding page.

	New Moon	First Quar.	Full Moon	Last Quar.
Rising				
Culmination				
Setting				

2. What phase occurs at position 4?

   What phase occurs at position 1?

   What phase occurs at position 2?

   What phase occurs at position 3?

3. What moon phase can create a **lunar eclipse**? _____ Sketch the moon, earth and sun in alignment and include the earth shadow and moon shadow. (Use the back of the page.)
4. What moon phase can create a **solar eclipse**? _____ Sketch the moon, earth and sun in alignment and include the earth shadow and moon shadow. (Use the back of the page.)
5. Why doesn't a lunar and solar eclipse occur every month?
6. What time does the Waxing Gibbous Moon **RISE**? (Use moon phase chart.) ____ Can you see the Waxing Gibbous Moon in **DAYLIGHT?**
7. You are planning a "star party" and the sky must be dark and "moonless" (when the moon is below the horizon between **sunset to midnight**). What moon phases will be ideal for this star party? _______________
8. At midnight where would the Last Quarter Moon be? __________
9. An occultation of Jupiter by the Waning Gibbous Moon will occur at 7:15 pm, local time. Can you see the event? (Where is the Waning Gibbous Moon at 7:15 pm?) ____

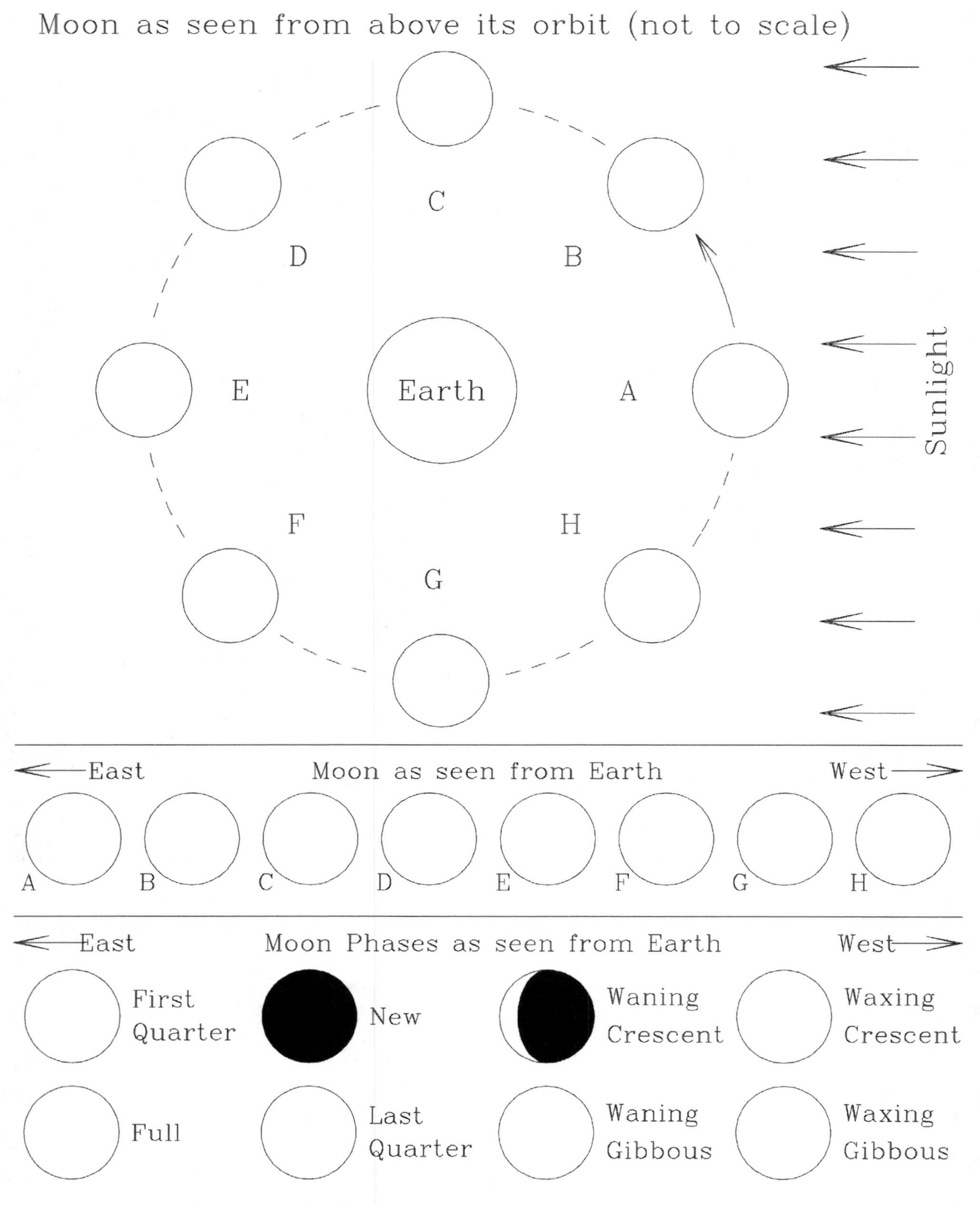

Figure 1: (Top) The Moon as seen from above its orbit. (Middle) The Moon as seen from Earth. (Bottom) Phases of the Moon. The Full, New, and Waning Crescent phases have been drawn in for you.

The Phases of Venus

TEAM #______________________________

Please print your name and sign next to it (only those present).

Leader: (C)______________________________ ______________________________

Explorer: (D)______________________________ ______________________________

Skeptic: (A)______________________________ ______________________________

Recorder: (B)______________________________ ______________________________

Learning Objectives

1. Use the Ptolemaic model to predict the phases of the Venus.
2. Use the Copernican model to predict the phases of the Venus.
3. Understand how Galileo's observation of the phases of Venus helps to select between these two models.

Introduction: As the first scientist to undertake detailed astronomical study with the aid of a telescope, Galileo Galilei (1564-1642) was able to make a number of important discoveries that greatly influenced our understanding of the universe. These observations were published in 1610 in his book *The Starry Messenger*, which was written in a style that made it accessible to a wide audience. One of his observations was that Venus exhibits phases much like our Moon.

Part I: Venus in the Ptolemaic Model

In the Ptolemaic model, Venus moves on an epicycle whose center is always directly between Earth and the Sun. The figure on page 35 shows Venus in a series of locations around its epicycle. Your task is to sketch the appearance of Venus (as observed through a telescope from Earth), when Venus is at each of the five indicated positions. Use the following procedure:

- For each of the five positions along the epicycle (i.e., the *top half* of the page) shade in the dark half of Venus.
- For each of these positions sketch (in the spaces at the *bottom* of the page) what Venus would look like as seen from Earth. In your sketches, shade the portion that would appear dark and the leave alone the part that would appear bright.

1. Is there any location along the epicycle at which Venus would appear as a near fully lit disk? (If there is, indicate which one.)

Part II: Venus in the Copernican Model

In the Copernican model, Venus moves around the Sun in an orbit that is only 72% the size of Earth's orbit. The figure on page 36 shows Venus at five different locations around the Sun. Your task is to repeat the procedure from Part I to sketch the appearance of Venus as seen from Earth when Venus is observed at each of these locations.

2. Is there any location along its orbit at which Venus would appear as a near fully lit disk? (If there is, indicate which one.)

3. Galileo observed that Venus goes through a complete set of phases including full phase. With which system of planetary motion, Ptolemaic or Copernican, is this observation consistent? Explain.

4. Do these observations of the phases of Venus necessarily confirm the heliocentric model? (Hint: think about Tycho's model in which the Sun orbits the Earth but Venus orbits the Sun.) Explain your answer.

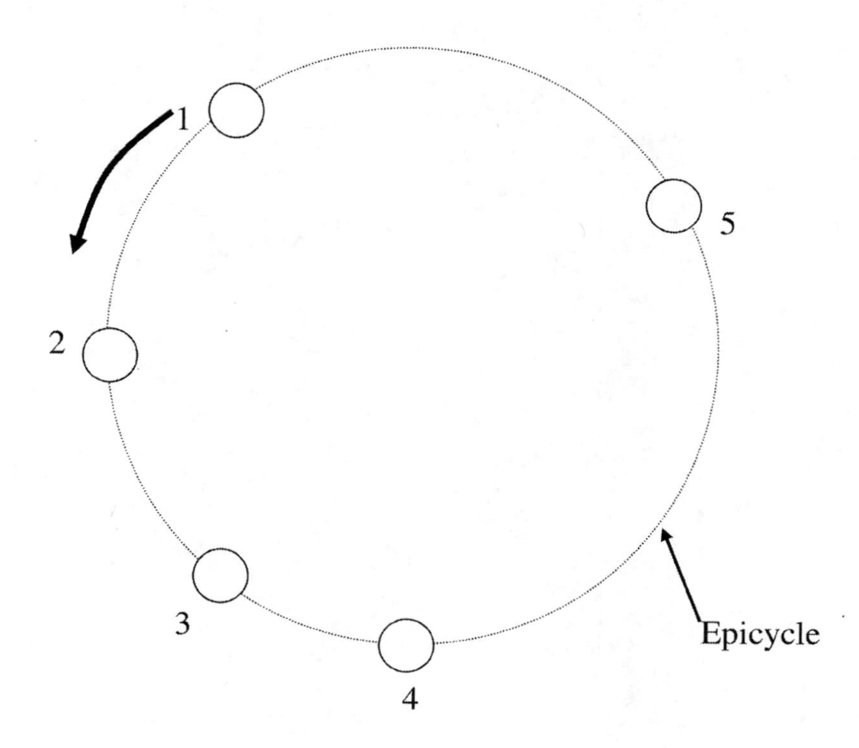

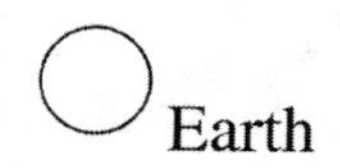

Sketch Venus's appearance as seen from Earth at the five locations shown above.

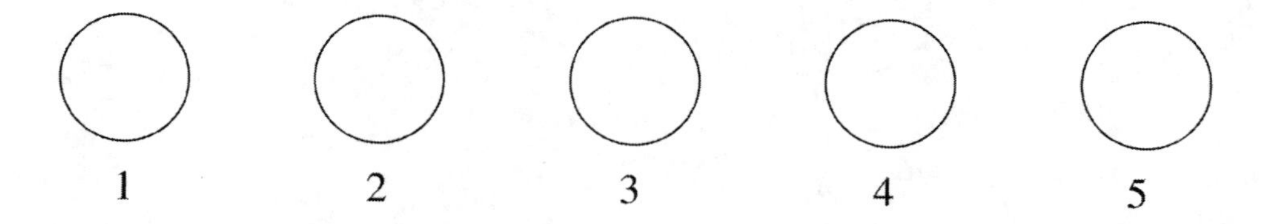

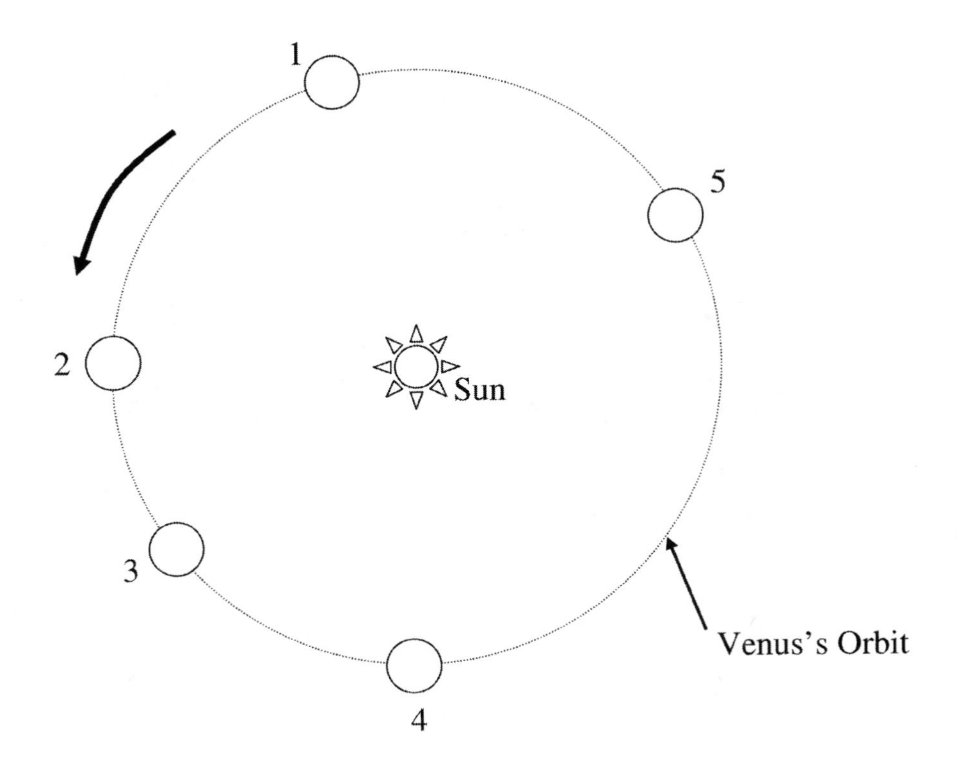

Sketch Venus's appearance as seen from Earth at the five locations shown above.

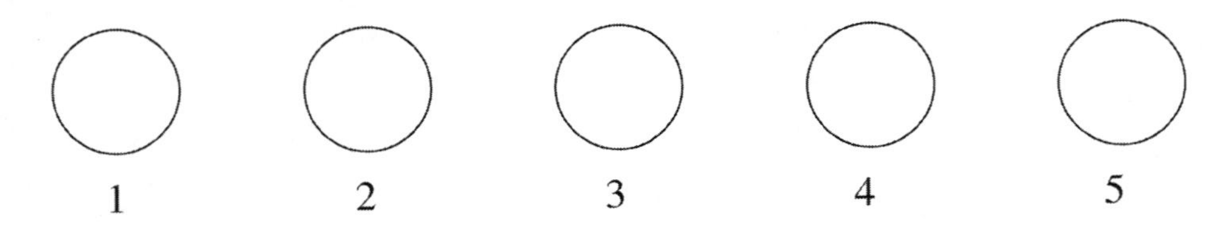

Mapping the Solar System from Earth

TEAM #______________________________

Please print your name and sign next to it (only those present).

Leader: (D)______________________________ ______________________________

Explorer: (A)______________________________ ______________________________

Skeptic: (B)______________________________ ______________________________

Recorder: (C)______________________________ ______________________________

Learning Objectives

1. Comprehend that the observer's position on Earth makes particular objects in the sky visible at specific times.
2. Analyze the rotation of an Earth observer to predict the rising & setting times of sky objects.
3. Synthesize heliocentric object locations and interpret to a geocentric perspective.
4. Synthesize geocentric object positions and interpret to a heliocentric perspective.

Background: Some newspapers and science magazines, such as *Sky and Telescope*, provide sky charts that describe what sky objects are visible at different times. These typically include

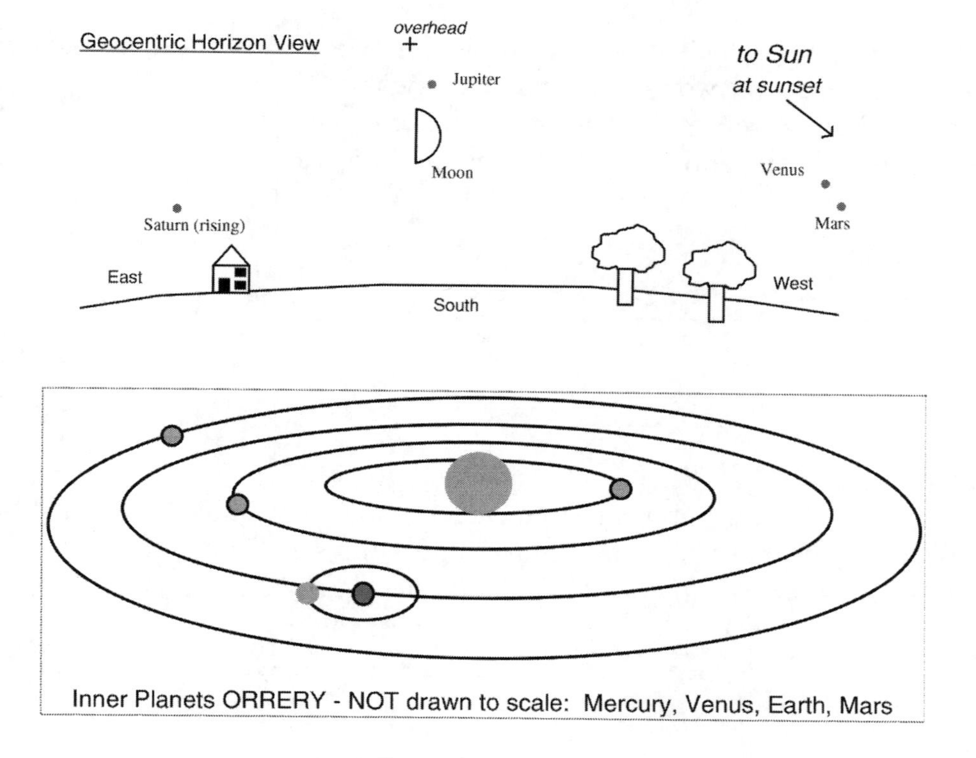

prominent stars, bright planets, and the Moon. There are two principle maps provided to readers: (1) a geocentric horizon view and (2) a heliocentric orrery view. The *geocentric* perspective is the view from Earth looking up into the southern sky. The *heliocentric* perspective is the view of the Solar System looking down from above. From above, the plants orbit and spin counter-clockwise (except Venus, which appears to spin backwards).

Part I: Rising and Setting Times

As seen from above, Earth appears to rotate counterclockwise. Figure I-a shows a top view of Earth and an observer at noon. Note that our Sun appears overhead when standing at the equator.

1. In Figure I-a, sketch and label the positions of the observer at midnight, 6 pm (sunset) and 6 am (sunrise).

Figure I-a: Observer Positions on Earth [Observer is at Equator]

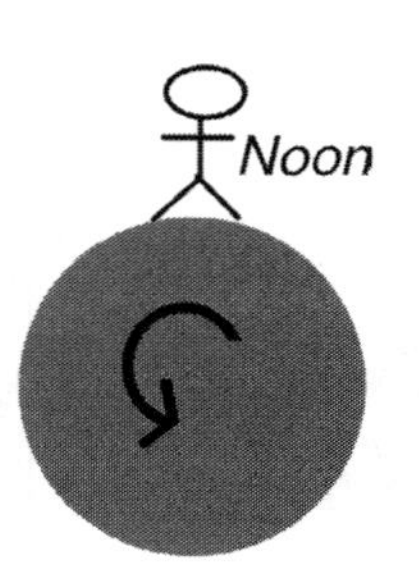

2. Consider Figure I-b, which shows Earth, Moon, Mars, and Venus. At what time would each of these sky objects be overhead? Remember that Earth spins counter-clockwise when viewed from above. [*Hint: Make use of Figure I-a*]

   <u>*Time Overhead:*</u>

   Venus: ____________________

   Moon: ____________________

   Mars: ____________________

Figure I-b: Orrery

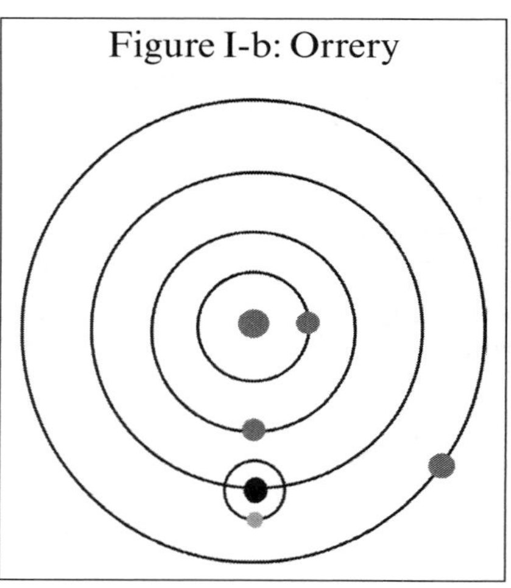

3. If Earth spins in 24 hours, it means that each sky object is visible for about 12 hours. What time will the sky objects shown in Figure I-b rise and set? Complete the table below? *Each member of your team should fill in the data for one sky object.*

Sky Object	Rise Time	Time Overhead	Set Time
Sun			
Venus			
Moon			
Mars			

4. Using complete sentences, explain why our Sun is not visible at midnight. Add a sketch of Earth, Sun, and observer in the space provided to support your explanation.

Narrative	Sketch

Part II: Converting Geocentric to Heliocentric

5. Figure II-a shows the horizon view of the first quarter Moon and Saturn visible at sunset. On the orrery shown in Figure II-b, sketch and label the position of Jupiter, Moon and Saturn. Use an arrow to indicate the direction to our Sun. Start by indicating the position of the observer at sunset. After completing the diagram, complete the table.

Sky Object	Rise Time	Set Time
Sun		
Jupiter		
Moon		
Saturn		

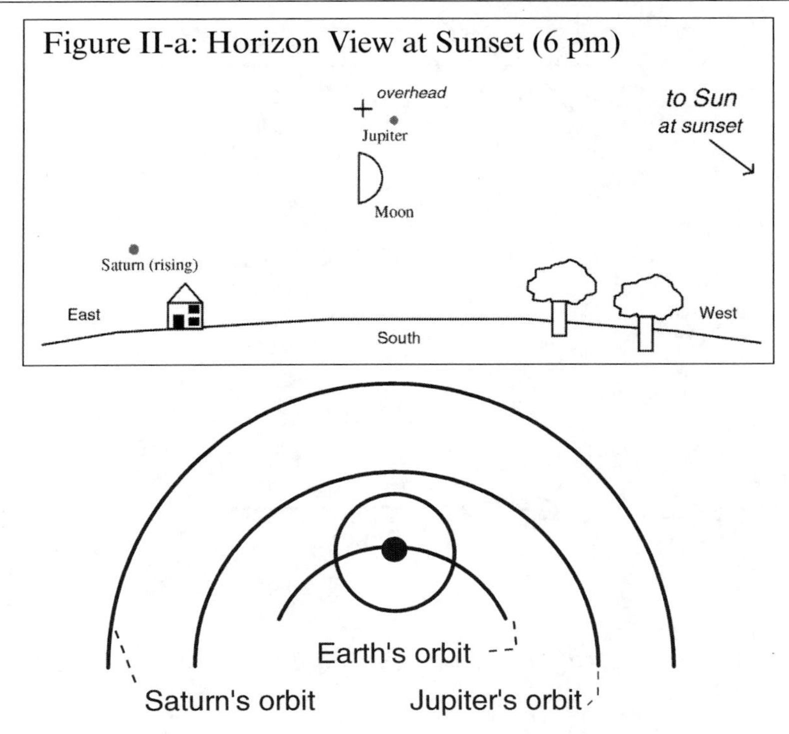

Figure II-a: Horizon View at Sunset (6 pm)

Orrery Not Drawn to Scale !!

Figure II-b

6. If Neptune is visible overhead in the southern sky at sunrise (6 am) sketch the relative positions of Sun, Earth, Neptune, and observer in an orrery in the space below.

Part III: Converting Heliocentric to Geocentric

7. Figure III-a shows the position of Mercury, Venus, Earth, Mars, and Moon. On the horizon diagram, Figure III-b, sketch and label the positions of Mercury, Venus, Mars, a comet, and Moon at midnight.

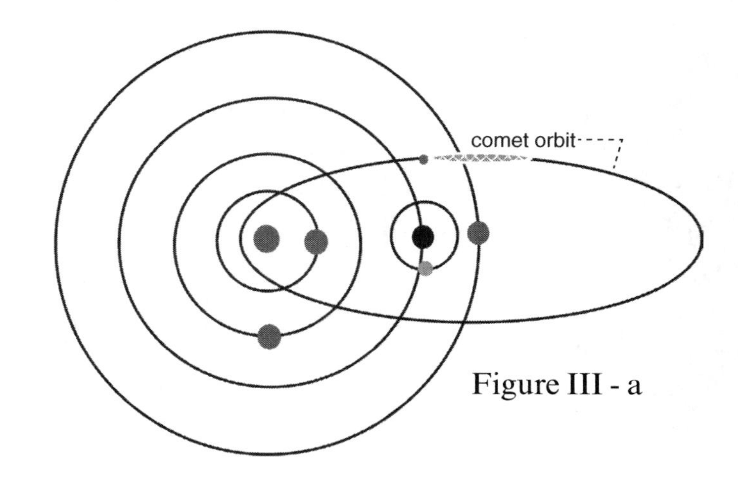

Figure III - a

Figure III-b: Geocentric Horizon View at Midnight

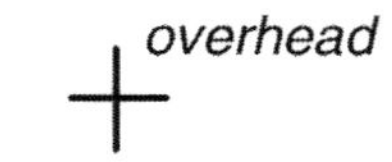

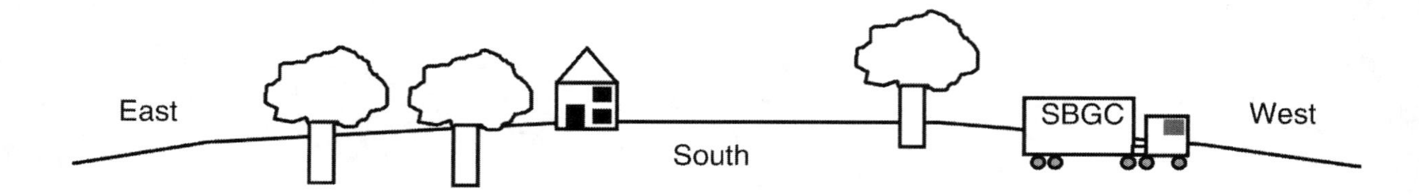

8. Venus is often called the *morning star* or the *evening star*. Why is it never seen at midnight?

Part IV: Current Events

9. Your instructor will provide you with a magazine photocopy or software print-out of the current sky or orrery. Convert the given current sky to an orrery OR convert the current orrery to a horizon view (you can select the time). Make your sketches in the space below. Be certain to label every item carefully.

10. Describe and sketch the night sky if you were to go outside at midnight tonight (*assume there are no clouds*).

Moon Phases

TEAM # ______

Please print your name and sign next to it (only those present).

Leader: (A)______________________ ______________________

Explorer: (B)______________________ ______________________

Skeptic: (C)______________________ ______________________

Recorder: (D)______________________ ______________________

Learning Objectives

1. Differentiate between looking from above the solar system versus standing outside.
2. Develop a mental model of the Sun-Earth-Moon geometry responsible for lunar phases.
3. Predict the phase of the Moon given the relative positions of the Sun-Earth-Moon system.
4. Predict the relative positions of the Sun-Earth-Moon system given the Moon phase and time of day.

Introduction: Throughout history, many cultures have used the lunar cycle as a method to measure increments of time—a lunar calendar. Nomadic peoples regulated their calendar based entirely on the 28-day cycle of the Moon. Every time the slender crescent after a new Moon appeared in the western evening sky, a new month began at the evening hour. The lunar cycle was quite practical because of its short duration and its ease in use. The phases of the Moon are due to the changing relative positions of the Sun, Moon, and Earth. One half of the Moon is always lit by the Sun. As the Moon orbits around the Earth, we see first more and more of the lit half and then, less and less.

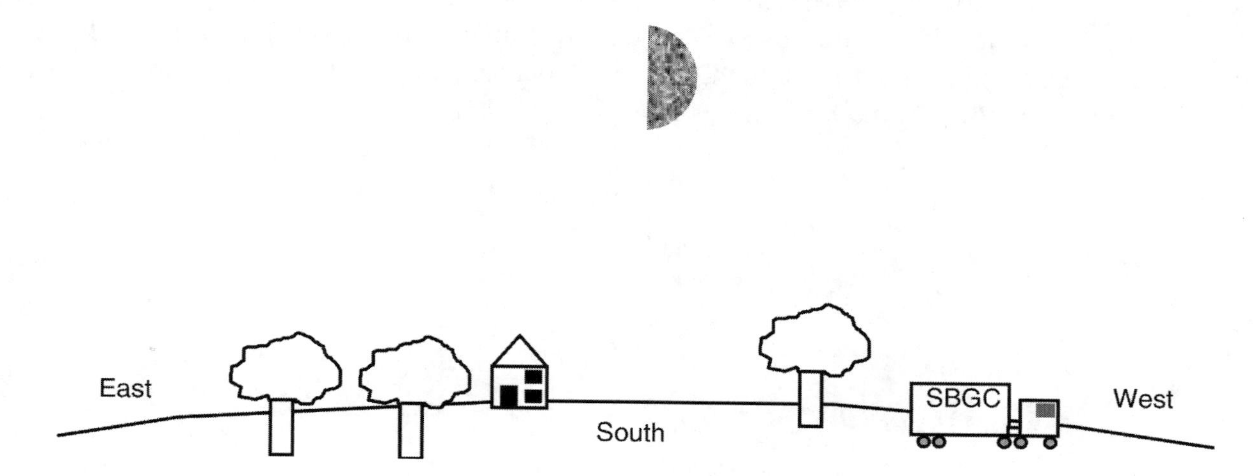

Figure III-b: Geocentric Horizon View at Midnight

For all of these Moon phase activities, assume that you are facing south. When the Moon is setting in the west it will be directly to your right. When it is rising in the east it will be directly to your left. When the Moon is straight up from the southern horizon—and fairly high overhead—we will refer to it as being south. *Questions marked with an * require you to use the above picture.*

Part I

The figure below shows an observer on Earth's equator as seen from above the north pole. The Earth appears to spin counterclockwise when viewed from above.

1. Draw and label the observer where she would be at midnight, sunset (6 PM), and sunrise (6 AM). For the 6 AM and 6 PM locations, draw and label arrows to indicate east and west. (Looking from *above* the solar system.)

Part II

Because in our everyday experience we tend to think about the Moon and Sun moving about a stationary Earth, we will now imagine looking down upon our observer from the above the north pole and spinning with the Earth. The figure below shows the direction from which the sunlight would be coming at noon for the observer positioned as shown.

2. *Draw arrows showing the direction from which the sunlight would be coming at sunset, midnight, and sunrise at the observer's location. Label your arrows accordingly. (Hint: if you are stuck, refer back to Part I.) Draw and label arrows to indicate east and west at the observer's location. (*Standing by the noodle, facing south.*)

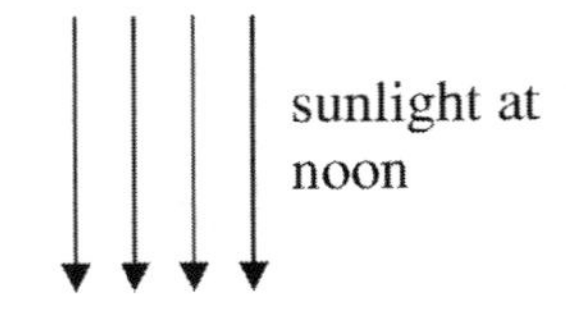

Part III

The figure below shows the Earth-Moon-Sun system as viewed from above. The Sun always illuminates the half of the Moon facing the Sun. It looks like the Earth might block the sunlight from reaching the Moon but this rarely happens because the Moon is normally above or below a straight line drawn from the Sun to the Earth as shown at right. Refer to Part I.

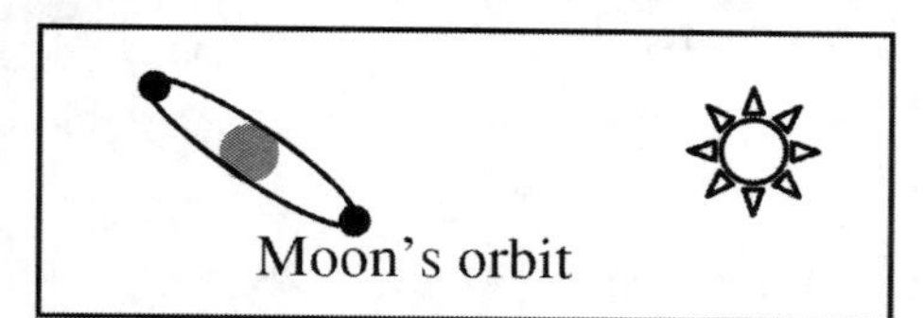

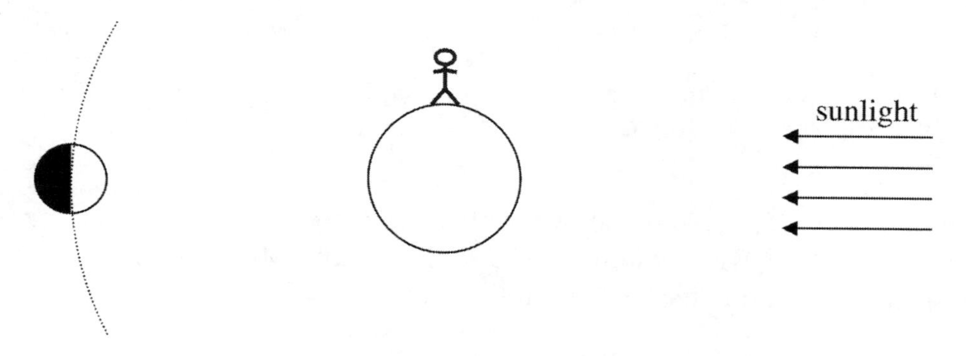

3. What time is it at the observer's location in the figure above?

4. *The observer looks to the west to see the Sun. What direction does she look to see the Moon?

5. *In the box at right, sketch what the observer would see looking at the Moon (shade in any portion that would not be visible). What do we call this phase (circle)? The last page of this activity is a *Moon Phase Poster*, giving the names of the different phases.

   new   *crescent*   *quarter*   *gibbous*   *full*

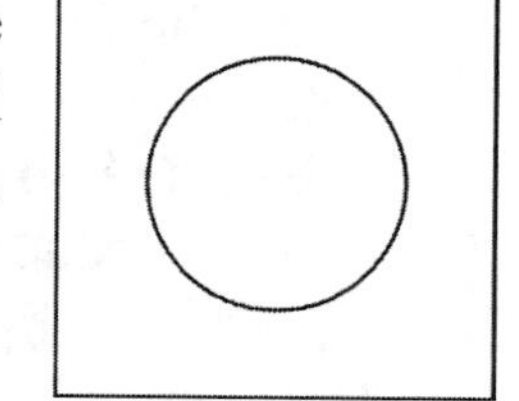

6. *At midnight, the Sun will be on the opposite side of the Earth from the observer and the Moon will be overhead. What phase will the Moon appear then (circle)?

   new   *crescent*   *quarter*   *gibbous*   *full*

7. *In what direction would the observer look to see the Moon six hours later at sunrise?

8. What phase will the Moon appear at sunrise (circle)?

   new   *crescent*   *quarter*   *gibbous*   *full*

9. Will the Moon be visible at noon this same day?
   Explain your answer with a sketch at right.

Part IV

The Moon is in its new phase when the lit side is on the far side of the Moon from the observer. The figure below shows the observer at sunset with the Sun in the west.

10. *Sketch the Moon on the figure above so that the observer would say it is in its new phase. As was done in Part III, shade in the dark side of the Moon to show which side of the Moon is actually illuminated by the Sun.

11. At what time will the Moon set below the horizon?

12. At what time will the Moon rise <u>and</u> in what direction would you look to "see" it (assuming it could actually be seen)?

Part V

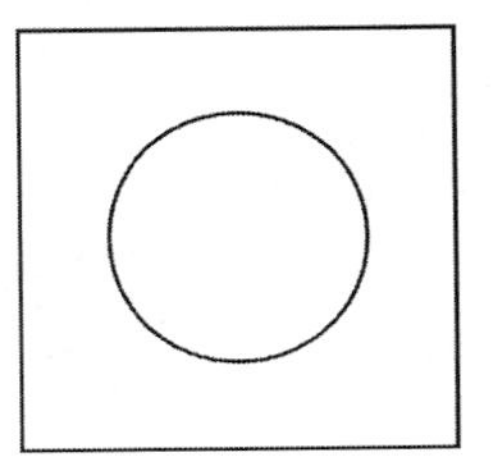

13. A few days later, the Moon will be about half way between overhead and the western horizon as the Sun is setting in the west as shown on the figure below. Again, the side illuminated by the Sun is shown. Remember that the Sun is very far away. In the box at right, sketch what the observer would see looking at the Moon (shade in any portion that would not be visible). What do we call this phase (circle)?

    new crescent quarter gibbous full

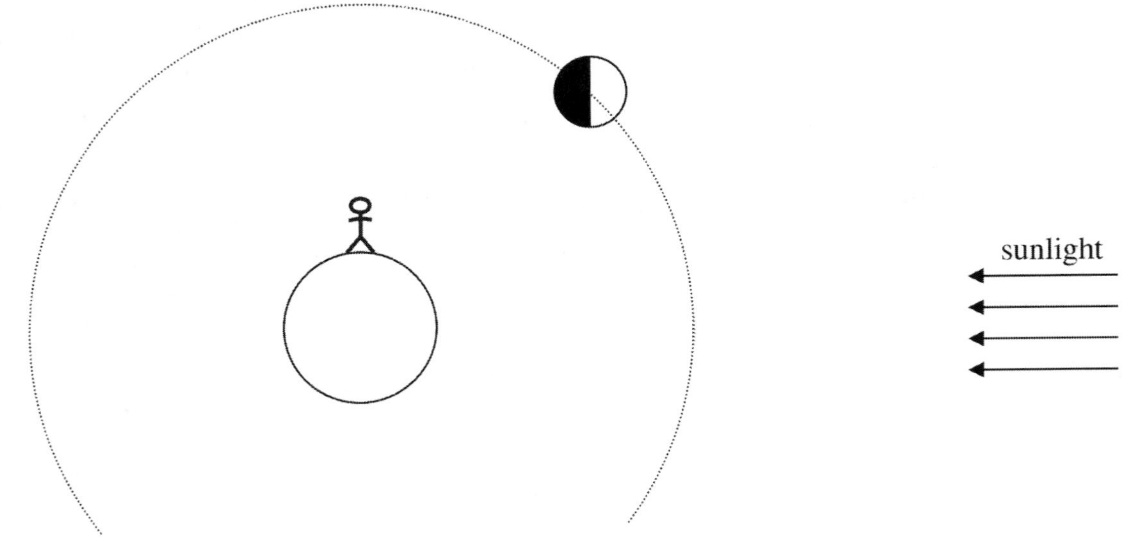

14. At about what time will this Moon set?

15. At about what time will this Moon rise?

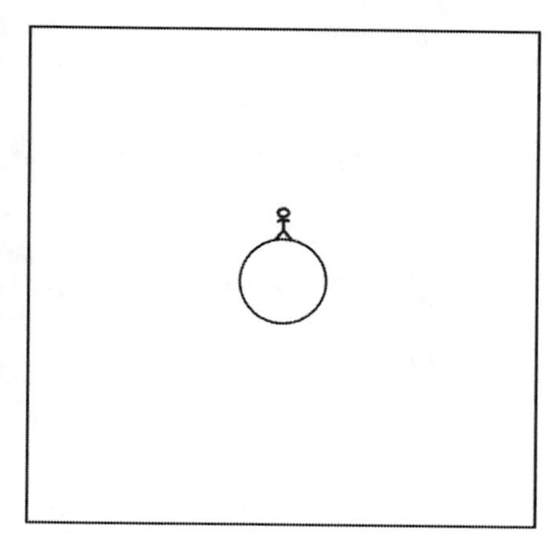

16. *In the box at right, locate the Moon and the direction of the sunlight as this Moon is rising.

17. *In the Northern Hemisphere, we often say that the Sun is in the south when at its highest point, which we have called overhead until now. We can then describe the Sun's position changing throughout the day with the sequence east (6 AM), southeast (9 AM), south (noon), southwest (3 PM), and west (6 PM). In what direction does the observer look to see the Sun when the crescent Moon is rising (circle)?

   east *southeast* *south* *southwest* *west*

18. In what direction would the observer look to see the crescent Moon at noon (circle)?

   east *southeast* *south* *southwest* *west*

19. If the observer were looking toward the Moon at noon, would she see the Moon's illuminated crescent on the right side or the left side (circle)? *right* *left*

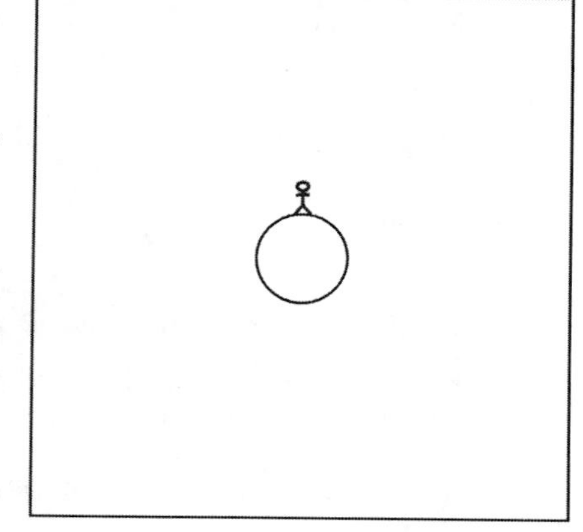

Part VI

About one week after the new Moon, the Moon will be to the south (overhead) at sunset.

20. In the box at right, locate the Moon and the direction of the sunlight at this time. Shade in the dark side of the Moon to show which side of the Moon is actually illuminated by the Sun.

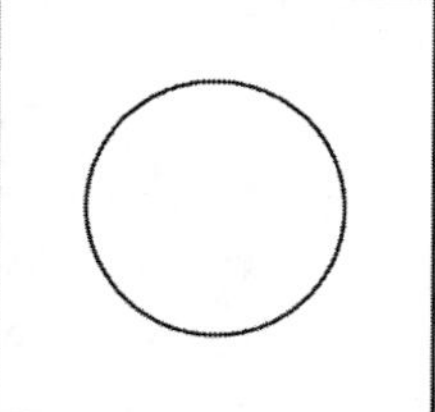

21. In the box at right, sketch what the observer would see looking at the Moon (shade in any portion that would not be visible). What do we call this phase (circle)?

   new *crescent* *quarter* *gibbous* *full*

22. Because the Moon can take on this same basic appearance (half lit, half dark), we differentiate between a first quarter Moon, which occurs about one week after the new Moon and a third quarter Moon, which occurs about three weeks after the new Moon. If the observer were looking toward a first quarter Moon at sunset, would she see the Moon's illuminated portion on the right side or the left side (circle)?

   right *left*

23. *In the box at right, indicate and label the locations of the first quarter Moon at each of the times below. In cases where the Moon is above the horizon, describe its direction as east, southeast, south, southwest, or west. If it is below the horizon, state that explicitly. The first one is done for you.

    6 PM: *south*

    9 PM:

    Midnight:

    6 AM:

    Noon:

    3 PM:

Part VII

About a week and a half after a new Moon, the Moon will be to the southeast (half way between the eastern horizon and overhead) at sunset.

24. *In the box at right, locate the Moon and the direction of the sunlight at this time. Shade in the dark side of the Moon to show which side of the Moon is actually illuminated by the Sun.

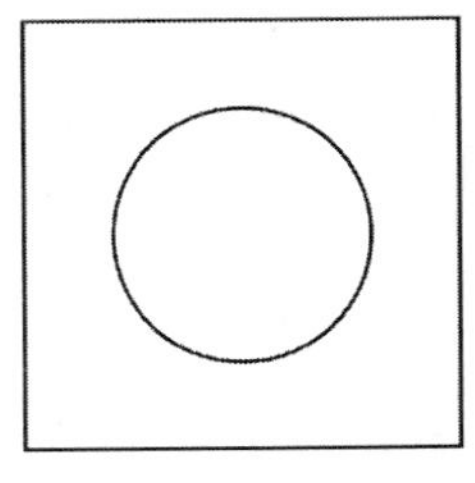

25. In the box at right, sketch what the observer would see looking at the Moon (shade in any portion that would not be visible). What do we call this phase (circle)?

    new *crescent* *quarter* *gibbous* *full*

26. Any time the Moon's lit portion is growing from one night to the next we say that it is **waxing**. When the Moon's lit portion is shrinking we say that it is **waning**. Is the Moon in this part waxing or waning? Explain how you know.

27. Imagine that you see a waxing crescent Moon setting on the western horizon. In the box at right, locate the Moon and the direction of the sunlight at this time. Shade in the dark side of the Moon to show which side of the Moon is actually illuminated by the Sun. About what time is it?

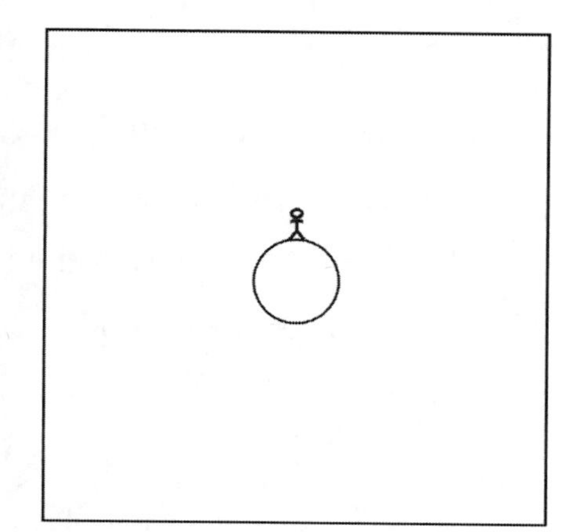

Part VIII

The third quarter Moon occurs about 3 weeks after new Moon. Whereas a first quarter Moon appears lit on the right side, a third quarter Moon appears lit on the left side when in the south (overhead).

28. *In the box at right, locate the Moon and the direction of the sunlight when the Moon is overhead. Shade in the dark side of the Moon to show which side of the Moon is actually illuminated by the Sun. About what time is it?

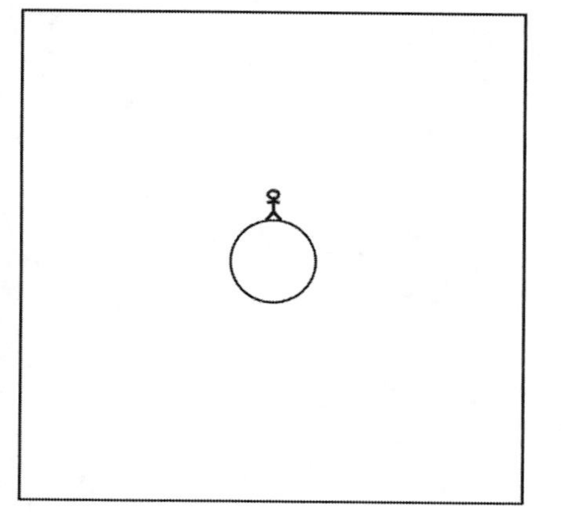

29. At what time does the third quarter Moon rise? Explain your reasoning.

30. At what time does the third quarter Moon set? Explain your reasoning.

31. You see a third quarter Moon in the southwest sky. In the box at right, locate the Moon and the direction of the sunlight at this time. Shade in the dark side of the Moon to show which side of the Moon is actually illuminated by the Sun. About what time is it?

Part IX: Putting it all together

Work as a group to find the correct answer to each of the following.

32. Which one of the following is a possible observation at 4 a.m.?
    a. a full Moon just rising in the east
    b. a crescent Moon just setting in the west
    c. a crescent Moon rising in the east
    d. a gibbous Moon rising in the east
    e. a first-quarter Moon in the south-west sky
33. An astronaut is standing on the Moon when the Moon is in a crescent phase. In what phase does the Earth appear to the astronaut? It will help to draw a diagram.
    a. full
    b. gibbous
    c. quarter
    d. crescent
    e. new

Moon Phase Poster

	New Moon	Day 0 and Day 28
	Waxing Crescent	Days 1-6
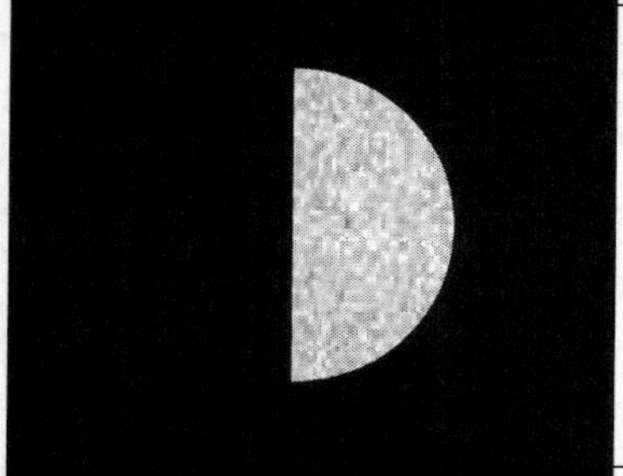	First Quarter	Day 7
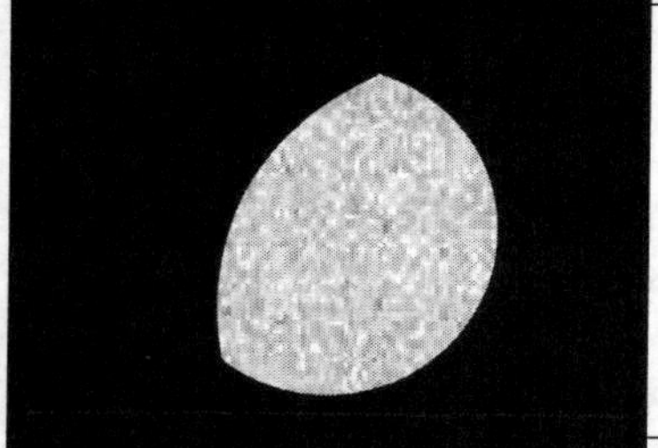	Waxing Gibbous	Days 8-13
	Full Moon	Day 14
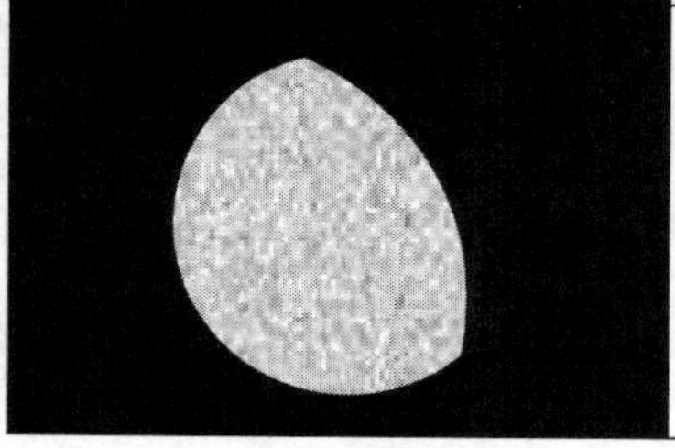	Waning Gibbous	Days 15-20

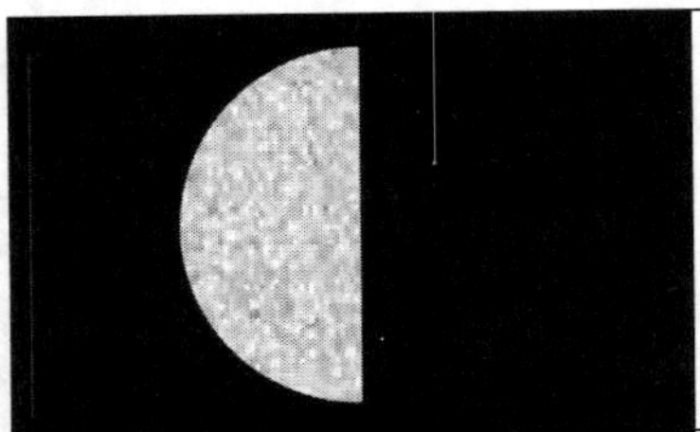	Third Quarter	Day 21
	Waning Crescent	Days 22-27

The Seasons and the Sun's Path on the Celestial Sphere

NAME ______________________________ ID# ______________________________

DATE ______________________________

Figure 1 on the next page shows the visible hemisphere of the celestial sphere for someone located at latitude 45 degrees north of the equator. If you are the stick figure, then the grey area is the part of the Earth's surface visible to you; the grey area ends at **your horizon.** The **cardinal directions** N(orth), E(ast), S(outh), and W(est) are indicated on your horizon.

The Sun's path through the sky on the **equinoxes** is drawn as the long-dashed line, passing through point b. The Sun's paths on the **solstices** are drawn as the two short-dashed lines, one passing through point a and the other through point c. Ignore the dotted line for now.

1. Mark your horizon and your **zenith** on Figure 1.
2. The Sun is shown crossing the meridian on a specific day of the year. What day is that? ______________ Label that position of the Sun on Figure 1 with that date.
3. Draw the Sun crossing the meridian on the remaining solstices and equinoxes, and label each position of the Sun with the date it would be seen there.
4. On what days does the Sun rise due East and set due West? ____________
5. Is the Sun ever directly overhead at this person's location? If so, on what day(s)?

   ____

   The following two questions refer to the Northern Hemisphere only.

6. Between what days does the Sun rise north of East and set north of West? What season(s) occur at that time of year? __________________
7. In what season(s) does the Sun rise south of East and set south of West? ____________

Figure 2 shows the southern half of the sky visible from the position of the stick figure in Figure 1. The **visible** celestial hemisphere has been cut in half on the dotted line (which is the same in both figures) and its southern half is shown. The Earth (grey) ends at the horizon, and the zenith is marked. The long-dashed and short-dashed lines are the same as in Figure 1.

8. Draw in the part of your meridian visible in Figure 2.
9. Add the labels a, b, and c to the appropriate locations in Figure 2.
10. Shade in the parts of the sky in Figure 2 in which the Sun will never be seen.
11. **(Optional)** If you observe the Sun move along the dotted line in Figure 1 over the course of a day, where on Earth are you located, and what day of the year is it?

    ____

12. **(Optional)** What is the maximum altitude reached by the Sun on an equinox, and how does this relate to the latitude of the stick figure? (Hint: The zenith is 90° from the horizon.)

    ____

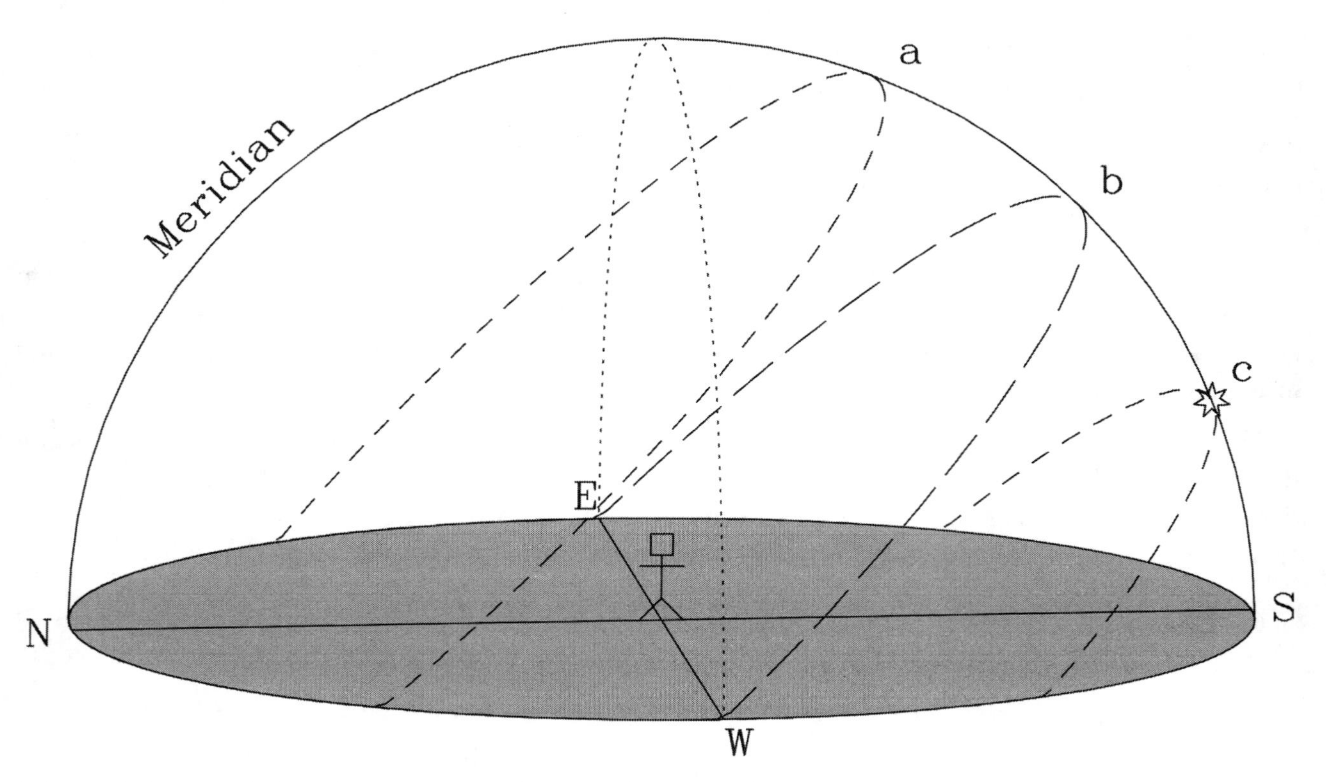

Figure 1: The visible celestial sphere from latitude 45 degrees north of the equator. You are the stick figure, and the part of the Earth you can see (grey) ends at your horizon. The Sun (star) follows the long-dashed line on the equinoxes and the short-dashed lines on the solstices.

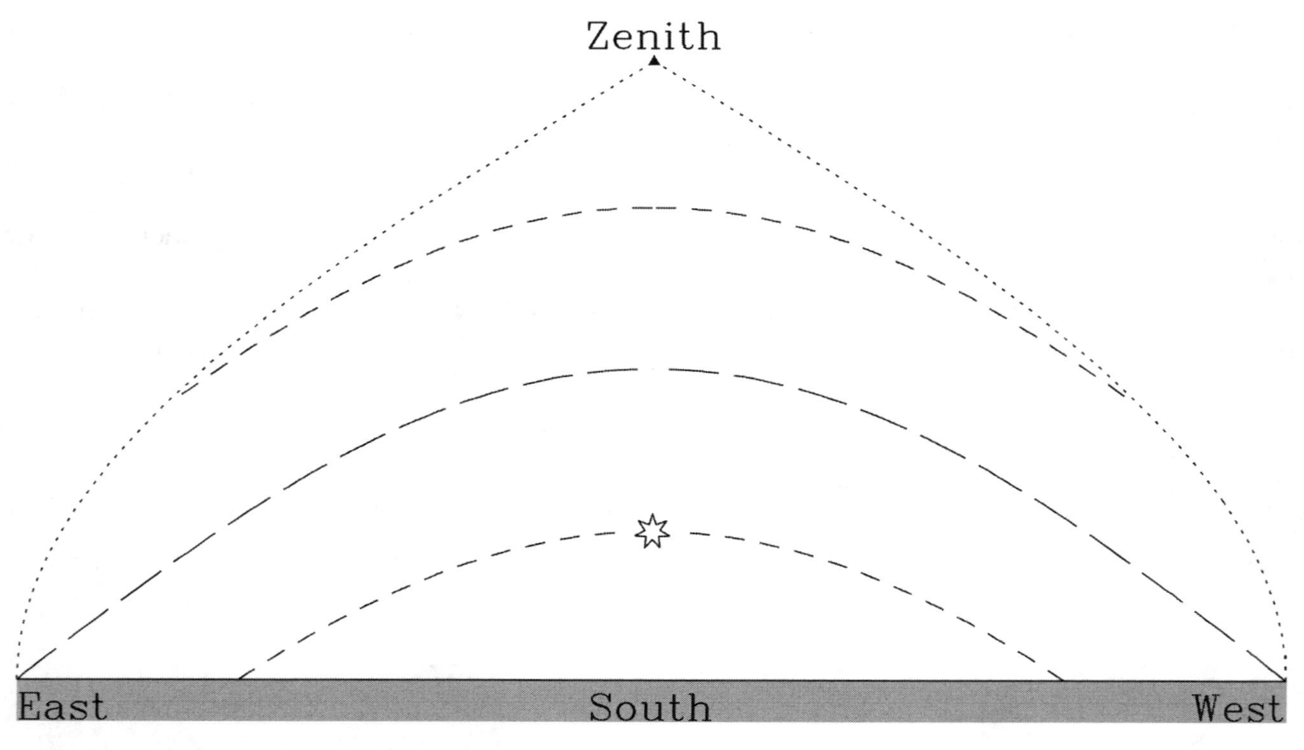

Figure 2: The southern half of the sky visible from the position of the stick figure in Figure 1. The visible celestial sphere in Figure 1 has been cut on the dotted line and the southern half of the sky is shown here. The lines and Sun symbol are the same as in Figure 1.

Force, Acceleration, and Gravity

NAME ______________ **ID#** __________________ **DATE** _________________________

Newton's Second Law of Motion

Isaac Newton published his **Three Laws of Motion** in 1687. To a very good approximation, these laws describe the motion of all objects in the universe.

Newton's Second Law of Motion can be stated as:

$$\text{Acceleration of object } (a) = \frac{\text{Force applied to object } (F)}{\text{mass of object } (m)}$$

Force is how hard something pushes or pulls on another thing. Acceleration is **how fast velocity (v) changes with time**. For example, if you are in a car that goes from 0 to 100 kph in 10 seconds, you experienced an acceleration of 10 kph per second.

1. Suppose you kick a football as hard as you can and then kick a sandbag of the same size with the same force.

   How far will the football go? ______________________________

   How far will the sandbag go relative to the football? ______________________________

2. Let's compare your common sense to Newton's Second Law. If the sandbag has 50 times the **mass** (m) of the football, which object will have a larger acceleration? ______________

3. Your foot touches the football and the sandbag for the same length of time (t). The velocity you give each object is, therefore:

   $$v = at.$$

   How much greater velocity will the football have than the sandbag? ______________

Newton's Universal Law of Gravitation

Newton worked out that the force of gravity F_g between two objects with masses m_1 and m_2 depends on the product of those masses divided by the square of the distance d between them:

$$F_g = \frac{Gm_1m_2}{d^2}$$

where G is a **constant** whose numerical value depends on the units you use. (If you measure the force between two specific objects in **Earth-pounds**, 1 Earth-pound being the force needed to hold up a weight of 1 pound in Earth's gravity, you must get the same answer for F_g in Earth-pounds no matter what units you use for mass and distance.) We will work in units where $G = 1$.

Figure 1 shows four pairs of asteroids with masses and distances given in units where $G = 1$. The asteroids have been held in place until this instant, so they have zero velocity. Each asteroid feels a gravitational force (and thus an acceleration) only from the other member of the pair.

4. **Estimate** or **guess** which pair of asteroids in Figure 1 experiences the strongest gravitational force? (Don't do any calculations!) ________________________

5. **Calculate** the gravitational forces between the pairs of asteroids in Figure 1. For example, if there was a pair of asteroids labeled Z with $m_1 = 3$, $m_2 = 2$, and $d = 1$, the force between those two asteroids would be $F_Z = (m_1 \times m_2)/d^2 = (3 \times 2)/1^2 = 6$ units.

   $F_A =$ $\quad$ $F_B =$ $\quad$ $F_C =$ $\quad$ $F_D =$

6. Use those results to complete this sentence: The weakest gravitational force is felt by ________, the next weakest by _______, the second strongest by _______, and the strongest by ________.

Acceleration Due to Gravity

We've worked out the gravitational force acting between each pair of asteroids, but what we observe is the acceleration of each asteroid. We can combine Newton's Second Law of Motion and Newton's Law of Universal Gravitation to find the acceleration of each asteroid in a pair:

$$a_1 = \frac{F_g}{m_1} = \frac{Gm_1m_2}{m_1d^2} = \frac{Gm_2}{d^2} \quad \text{and} \quad a_2 = \frac{F_g}{m_2} = \frac{Gm_1m_2}{m_2d^2} = \frac{Gm_1}{d^2}$$

Note that the acceleration of asteroid 1 does not depend on the mass of asteroid 1 and that the acceleration of asteroid 2 does not depend on the mass of asteroid 2. **The acceleration of an object due to the gravity of a second object depends only on the mass of the second object and the distance between the objects.**

7. Which asteroid in Figure 1 do you think experiences the largest acceleration? ___________

8. Calculate the acceleration of every asteroid in Figure 1. Express your answer as fractions.

   A: $a_1 =$ $\quad$ B: $a_1 =$ $\quad$ C: $a_1 =$ $\quad$ D: $a_1 =$

   A: $a_2 =$ $\quad$ B: $a_2 =$ $\quad$ C: $a_2 =$ $\quad$ D: $a_2 =$

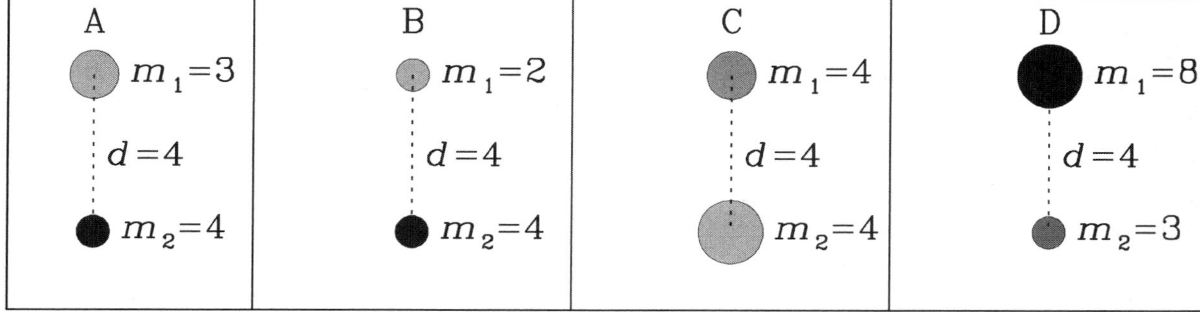

Figure 1: Four pairs of asteroids, with their masses and the distances between them given.

NAME ________________ ID# ______________ DATE ____________________

9. Which asteroid in Figure 1 has the smallest acceleration? ______________

   Which asteroid in Figure 1 has the largest acceleration? ______________

10. Was your answer to Question 7 correct? ______________

    Whether you were correct or not, explain how you would choose the correct answer next time:

    ____

    ____

11. Look at asteroids A1, B1, C1, and C2, and the asteroids each of them was paired with. What two things do asteroids A1, B1, C1, and C2 have in common? (Hint: The two things are related.)

    ____

    ____

12. If I replace asteroid A1, B1, C1, or C2 with Asteroid M, which has mass *M*, what will the acceleration of Asteroid M be, and how do you know?

    ____

    ____

13. Figure 2 shows four small asteroids next to one very large asteroid. Which small asteroid (A, B, C, or D) has the largest acceleration at the instant shown?

    A ________ B ________ C ________ D ________ All will have the same acceleration. ________

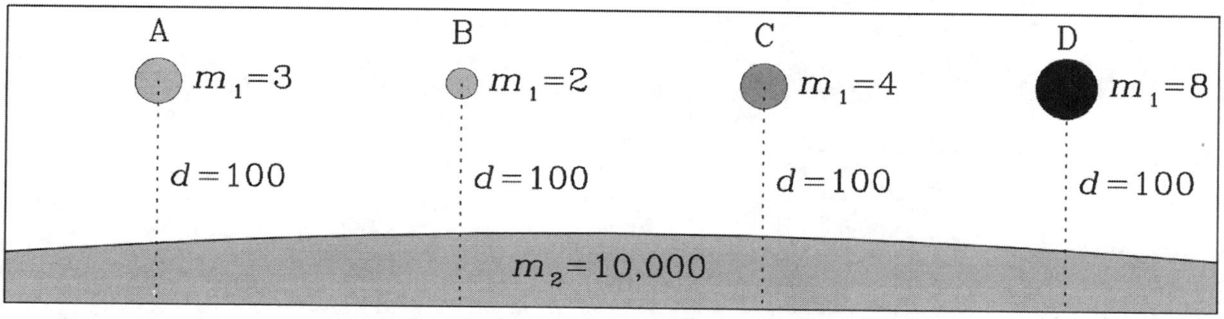

Figure 2: Four small objects next to one very large object.

14. Figure 2 could represent four spheres held above the surface of the Earth. The asteroid(s) or sphere(s) with the largest acceleration will be the first to collide with the object of mass m_2 (call it M2). (The objects are all the same distance from the surface and center of M2, and the acceleration of M2 is so small it can be ignored.) Based on your answer to Question 13, and on your experience here on Earth, which object(s) will collide with M2 first?

The following two questions summarize the difference between the **force of gravity** and the **acceleration due to gravity**.

15. The Earth pulls on you with a gravitational force which is... (choose one)
    a. larger than the gravitational force with which you pull on the Earth.
    b. equal to the gravitational force with which you pull on the Earth.
    c. smaller than the gravitational force with which you pull on the Earth.

16. Because of the force of gravity between you and the Earth, if you step off a table you will be accelerated toward the Earth. At the same time... (choose one)
    a. the Earth will be accelerated toward you much more than you are accelerated toward it.
    b. the Earth's acceleration toward you will exactly equal your acceleration toward it.
    c. the Earth will be accelerated toward you much less than you are accelerated toward it.

17. **(Optional)** Each asteroid pair in Figure 1 was held in place with rockets until this instant. Gravity will pull each pair of asteroids together. Which asteroid pair will collide first, and why?

    ____

    ____

Kepler's Laws and Elliptical Orbits

NAME ______________________ ID# ______________________

DATE ______________________

Introduction

Tycho Brahe observed the regularly changing positions of the planets relative to the stars over several decades. **Johannes Kepler** used these observations to determine how objects orbit the Sun and summarized his results in three Laws. Newton and Einstein later showed that these Laws are only approximations, but they are accurate enough for us to use in many cases.

Kepler's First Law

An object which orbits the Sun moves along an ellipse with the Sun at one focus.

An ellipse can be thought of as a stretched circle. An ellipse has two **foci** (the plural of focus) which are located inside the ellipse on its **major axis** (the longest direction of the ellipse). The **center of the ellipse** is exactly halfway between the foci.

An ellipse's **eccentricity** (e) is the distance between its foci divided by the length of its major axis. An ellipse with $e = 0$ is just a circle. An ellipse with $e = 1$ is infinitely long and thin.

The **semi-major axis** (a) is half the length of the major axis. For a circle, a equals the radius (r).

Use the above information to answer the following questions for the elliptical orbit in Figure 1.

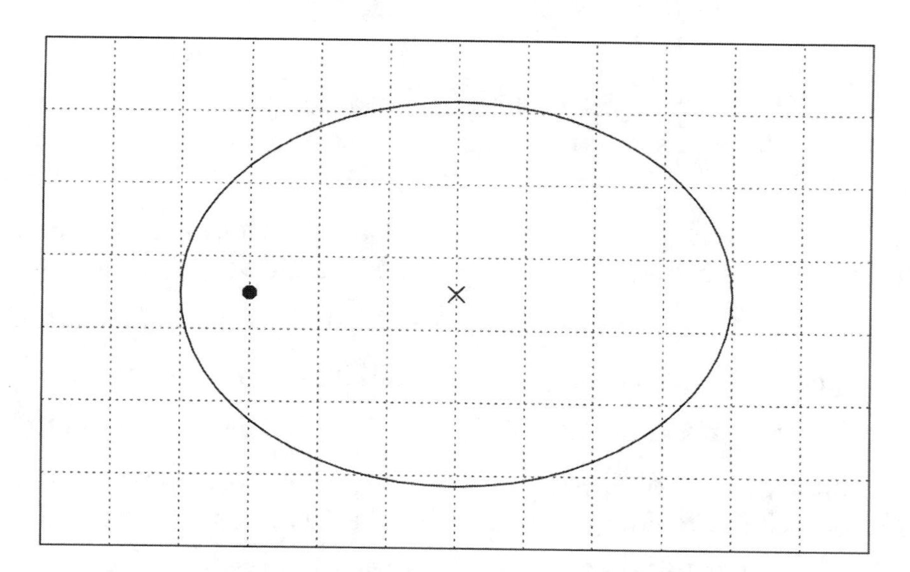

Figure 1: An elliptical orbit. The cross shows the center of the ellipse. Each dotted square measures 1 **astronomical unit** (AU) on a side.

1. One focus of the ellipse is shown by a black dot. Draw the Sun at the other focus.
2. How many AU across is the major axis of this ellipse? ______________________
3. The semi-major axis of this ellipse equals how many AU? ______________________
4. What is the eccentricity of this ellipse? ______________________

Kepler's Second Law

An object moves faster in its orbit when it is closer to the Sun, so that the line connecting the object and the Sun sweeps out equal areas in equal times.

Note that Kepler's Three Laws apply to all objects orbiting the Sun, not just planets.

The point in an object's orbit where it is closest to the Sun, at radius r_p, is called **perihelion**. The point where it is farthest from the Sun, at radius r_a, is called **aphelion**. To sweep out equal areas in equal times, the object moves $\frac{r_a}{r_p}$ times faster at perihelion than at aphelion. Perihelion and aphelion occur on the major axis, so $r_a + r_p = 2a$, where a is the semi-major axis.

5. Mark the perihelion of the orbit in Figure 1 with a P and the aphelion with an A.
6. How much faster is the object moving at perihelion than at aphelion? ______________

Kepler's Third Law

For objects orbiting the Sun, the square of the object's orbital period *P*, measured in years, equals the cube of its semi-major axis *a*, measured in astronomical units:

$$P^2 = a^3$$

The **orbital period** (P) of an object is how long it takes to move around its orbit once. The Earth has $P = 1$ Earth year and semi-major axis $a = 1$ astronomical unit (AU) by definition. For Earth, Kepler's Third Law $P^2 = a^3$ gives $1^2 = 1^3$ or $(1 \times 1) = (1 \times 1 \times 1)$, which is correct.

The orbits of five asteroids (labeled A through E) are shown in Figure 2. Some of the orbits are circular and some are elliptical. Refer to Figure 2 when answering the following questions.

7. What is the perihelion of each orbit, in AU? Fill in these values in the table.
8. What is the aphelion of each orbit, in AU? Fill in these values in the table.
9. What is the semi-major axis of each orbit, in AU? Fill in these values in the table.
10. Rank the orbital periods of the asteroids **from shortest to longest**. Exact values are not needed, just the **relative rankings**. Fill in these relative rankings in the table.

As an example of this kind of ranking, the eccentricities of each of the asteroid's orbits have been ranked and entered in the table. Three of the orbits are circular and, therefore, have zero eccentricity. Of the other two orbits, E is clearly more elliptical than D.

11. One of these asteroids is later discovered to be a comet. **Based just on their orbits,** which asteroid is it most likely to be, and why?

    ____

    ____

12. **(Optional)** What are the orbital periods of these asteroids, in Earth years?
13. **(Optional)** What are the exact eccentricities of the orbits of D and E?

NAME ______________________________ ID# ______________________________

DATE ______________________________

Asteroid Name:	A	B	C	D	E
Perihelion in AU					
Aphelion in AU					
Semi-Major Axis in AU					
Eccentricity	Zero	Zero	Zero	Low	Medium
Orbital Period Ranking					
Orbital Period (Bonus)					

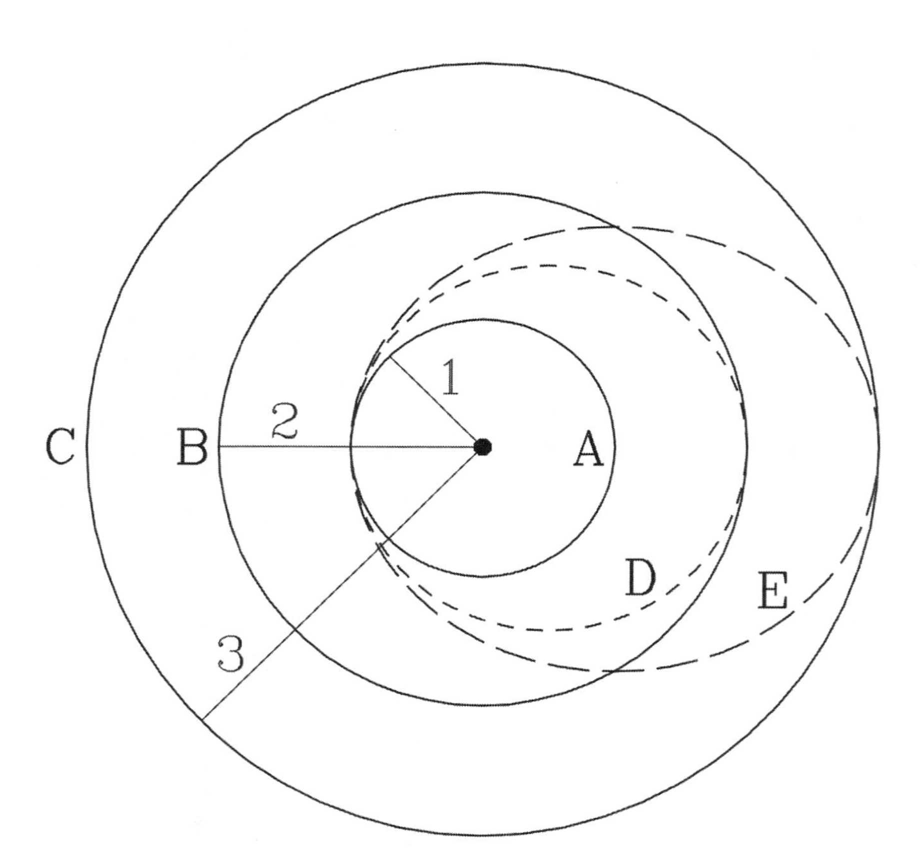

Figure 2: Circular and elliptical orbits of 5 asteroids (A through E) around the Sun (black dot). The numbers show the distances in AU to the circular orbits of asteroids A, B and C.

How Long Are the Days on Mercury and Venus?

NAME ____________________ ID# ____________________ DATE ____________________

A planet's **year** is how long it takes that planet to orbit the Sun.

A planet's **solar day**, usually just called its **day**, is the average time from noon to noon on that planet. In other words, it is the **synodic period** of the Sun as seen from that planet.

The length of a day and the number of days per year is different for each planet. For example, one **Mars day** is 1.027 Earth days long and one **Mars year** is 669 **Mars days** long (687 Earth days).

Let's work out how long the solar days are on the slowly spinning planets Mercury and Venus. To do this, we'll use our knowledge of how long the **sidereal days** are on Mercury and Venus. An object's sidereal day is how long it takes to spin completely around (360°) on its axis. We'll also need to know how long those planets take to orbit the Sun, and in what direction they spin and orbit.

Mercury

Figure 1 shows Mercury at six positions in its orbit around the Sun.

To a reasonable approximation, Mercury orbits the Sun counterclockwise once every 90 Earth days, and Mercury spins 360° counterclockwise on its axis once every 60 Earth days.

That is, the **sidereal day** on Mercury is about 60 Earth days long.

1. On Earth day 0, Mercury is at position A. What Earth day is it when Mercury is at positions B through G? (G is the same as A, but one orbit later.) Write those numbers on Figure 1.
2. At each position, draw in Mercury's **terminator** (the boundary between night and day). Then shade the night side of the planet. This has already been done for you at position F.
3. In both Figure 1 and Figure 2, in position A we have drawn a mountain on Mercury (call it Triangle Mountain).

   You know how long Mercury takes to spin around 360 degrees, so at each position in Figure 2 you can draw in Triangle Mountain.

   When you have filled in Figure 2, copy the location of Triangle mountain at each position onto the drawing of Mercury at the same position in Figure 1.
4. What time is it on Triangle Mountain when Mercury is at positions A, D, and G?

   Use terms like noon, early or late afternoon, sunset, midnight, etc.

   Once you have filled in positions A, D and G, you should be able to fill in the times for positions B, C, E and F. Position A has already been filled in for you:

   A: noon B: ________ C: ________ D: ________ E: ________ F: ________ G: ________

5. From the above, how many Mercury years is it between noon and midnight on Mercury?

   Therefore, how long is it from noon to noon on Mercury?

   Therefore, one **solar** day on Mercury is _____ Mercury years or _____ Earth days long.

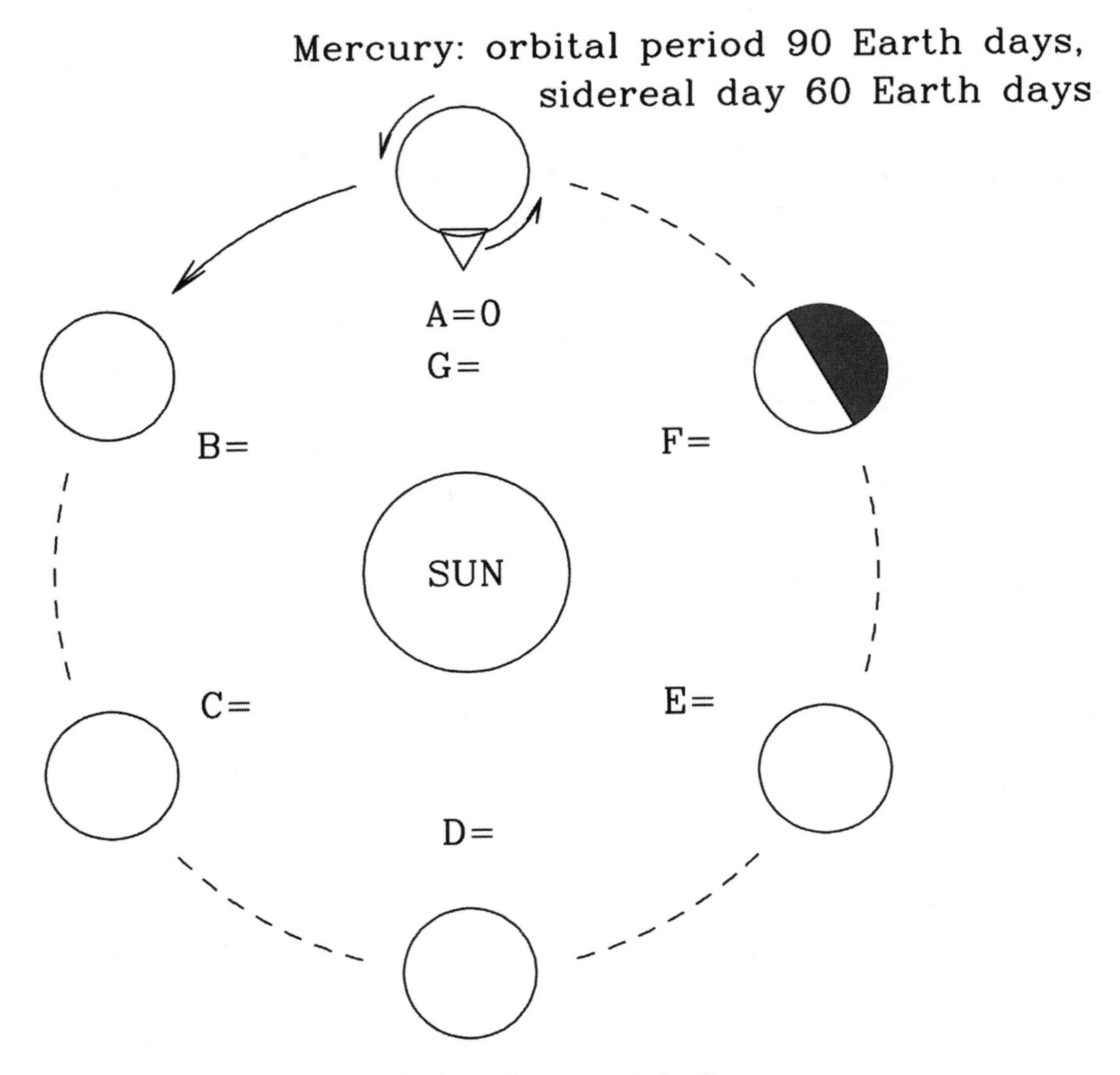

Figure 1: Mercury at six positions in its orbit around the Sun.

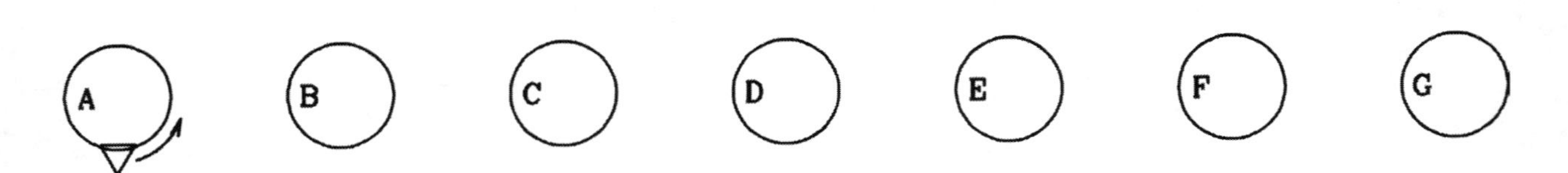

Figure 2: Mercury at the orbital positions of Figure 1. (Use this Figure for Question 3.)

Venus

In Figure 3, Venus is shown in eight positions in its orbit.

Venus orbits the Sun counterclockwise about once every 240 Earth days.

Venus spins 360° CLOCKWISE on its axis about once every 240 Earth days as well.

That is, the **sidereal day** on Venus is (about) 240 Earth days long.

6. At each position in its orbit, draw Venus' terminator, shade in its night side (see position F), and write in the number of the Earth day on which Venus will be at that position.

7. In both Figure 3 and Figure 4, we have drawn a mountain (Triangle Peak) on Venus in position A.

   You know how long Venus takes to spin around 360 degrees, so at each position in Figure 4 you can draw in Triangle Peak.

   When you have filled in Figure 4, copy the location of Triangle Peak at each position onto the drawing of Venus at the same position in Figure 3.

8. What time is it on Triangle Peak at each position in Venus's orbit? (Start with positions A, C, E, G, and I; Position I is the same as position A, but one orbit later.) Position A has already been filled in for you: A: noon

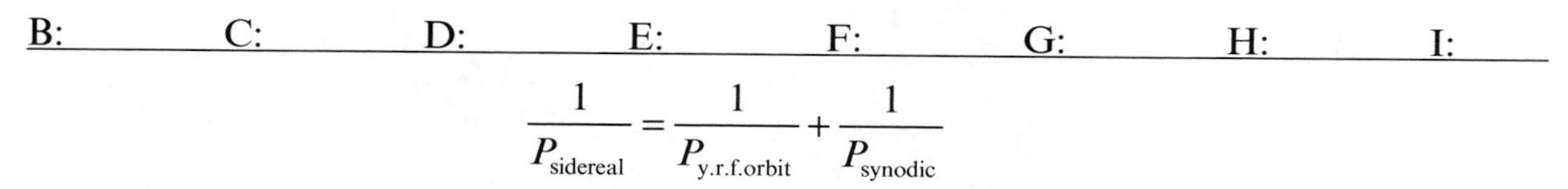

B: C: D: E: F: G: H: I:

$$\frac{1}{P_{\text{sidereal}}} = \frac{1}{P_{\text{y.r.f.orbit}}} + \frac{1}{P_{\text{synodic}}}$$

9. From the above, one **solar** day on Venus is _____ Venus years or _____ Earth days long.

10. Use the formula

    to confirm that the synodic periods of the Sun as seen from Mercury and Venus (the lengths of their solar days) match the results you obtained geometrically in questions 5 and 9.

11. Suppose an extrasolar planet orbits its star clockwise once every 4 days and also spins on it axis clockwise once every 4 days. Can you define the length of a solar day for this planet? If so, how long is a solar day on this planet?

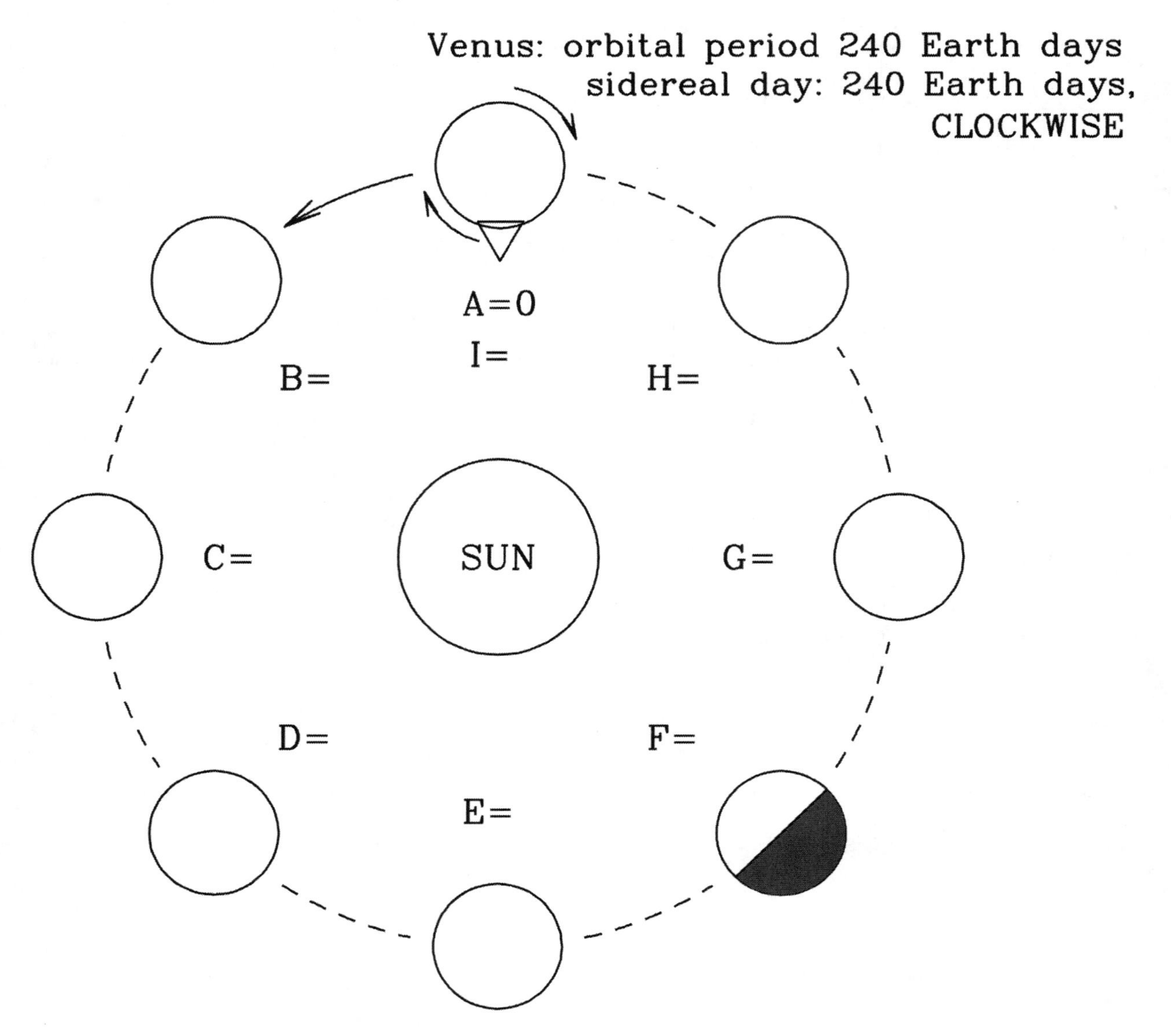

Figure 3: ABOVE: Venus at eight positions in its orbit around the Sun.

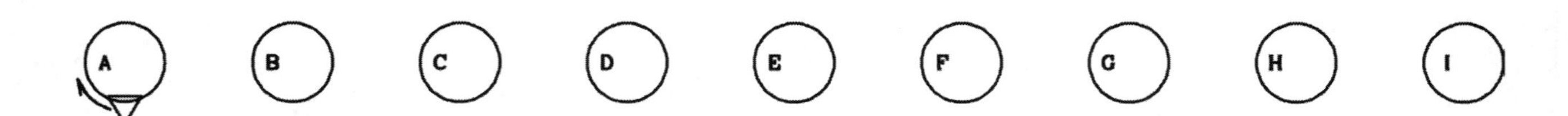

Figure 4: Venus at the orbital positions shown in Figure 3. The location of a mountain on Venus is shown in position A. Draw in the location of the mountain at positions B through I.

Electromagnetic Radiation and Thermal Spectra

NAME ________________ ID# ________________ DATE ________________

Any object can be characterized by a temperature. Temperature measures the average random motion or random vibration of an object's atoms or molecules, which include electrically charged electrons. Random motions or vibrations require accelerations to change directions, and accelerating charged particles emit **electromagnetic (EM) radiation**. Therefore, every **object emits some EM radiation.** EM radiation is often referred to as **light** by astronomers. EM radiation includes not just visible light but also gamma rays. infrared radiation (heat), radio waves, ultraviolet light, and X-rays.

Light consists of "particles" called **photons**. A photon has energy but no mass. Photons of any energy are possible. All photons travel at the speed of light.

A photon can also be thought of as an **electromagnetic wave**, with its **wavelength** (λ) and **frequency** (v) determined by its energy (E). Wavelength is the distance between wave peaks, often measured in **nanometers (nm)**: 1 nm = 1 billionth of a meter $\left(10^{-9}\text{ m}\right)$. Frequency is the number of times per second that a wave peak passes by you, and is measured in **hertz (Hz)**: 1 Hz = 1 peak per second. (Wavelength and frequency are related by the **speed of light** (c); for all photons, $c = \lambda v$.) The greater the photon energy, the more wave peaks pass by you per second, and the shorter the distance between them:

The frequency (v) of a photon is proportional to its energy: $v \propto E$

The wavelength (λ) of a photon is inversely proportional to its energy: $\lambda \propto \dfrac{1}{E}$

The above relationship means that if the energy of a photon doubles, its wavelength is halved.

1. Complete this sentence: If photon A has half the energy of photon B, then photon A will have ___________ the frequency of photon B and ___________the wavelength of photon B.

Thermal Spectra

The **luminosity** (L) of an object is the light energy per second emitted by that object. It can be measured over all wavelengths (**total luminosity**) or at each wavelength. When you measure an object's luminosity at each wavelength, you have measured that object's **spectrum**. Any sufficiently dense object will emit a particular kind of spectrum known as a **thermal spectrum**. The **shape** of this thermal spectrum depends **only** on the object's temperature, in two ways:

I. A hotter object emits more light per unit area at every wavelength than a cooler object.

II. A hotter object reaches its peak luminosity output per unit area and unit wavelength at a higher frequency (shorter wavelength) than a cooler object. In other words:

The frequency at which the peak luminosity output per unit area and unit wavelength of an object occurs is proportional to the object's temperature: $v_{peak} \propto T$

The wavelength at which the same peak occurs is inversely proportional to T: $\lambda_{peak} \propto \dfrac{1}{T}$

Figure 1 shows three thermal spectra (**spectra** is the plural of spectrum). The three objects plotted in Figure 1 have temperatures of 5850 K (the temperature of the Sun's surface), 3900 K (two-thirds the temperature of the Sun's surface), and 6435 K (ten percent hotter than the temperature of the Sun's surface).

2. The Sun emits a roughly thermal spectrum (we can ignore the Sun's narrow absorption lines in this exercise). According to Figure 1, at what frequency (ν_{peak}) does the Sun ($T = 5850$ K) emit the highest luminosity per unit area and unit wavelength? ____________________

3. The object with $T = 3900$ K has a surface temperature two-thirds of the Sun's. What should its ν_{peak} be, based on your answer to Question 2? ______________ What ν_{peak} do you measure for it in Figure 1? ______________ Are your answers in reasonable agreement? *(Yes | No)*

4. Visible light occupies the grey shaded region in Figure 1. When an object heats up to a sufficiently high temperature, it glows reddish, then yellowish, then bluish. Therefore, which photons have higher energy: red photons or blue photons? ______________ Based on that, correctly label the ends of the grey shaded region in Figure 1 with "blue" and "red".

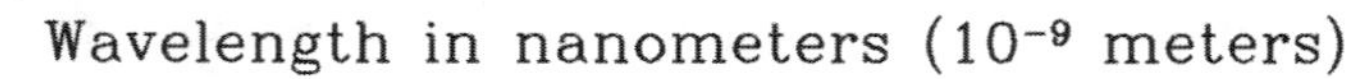

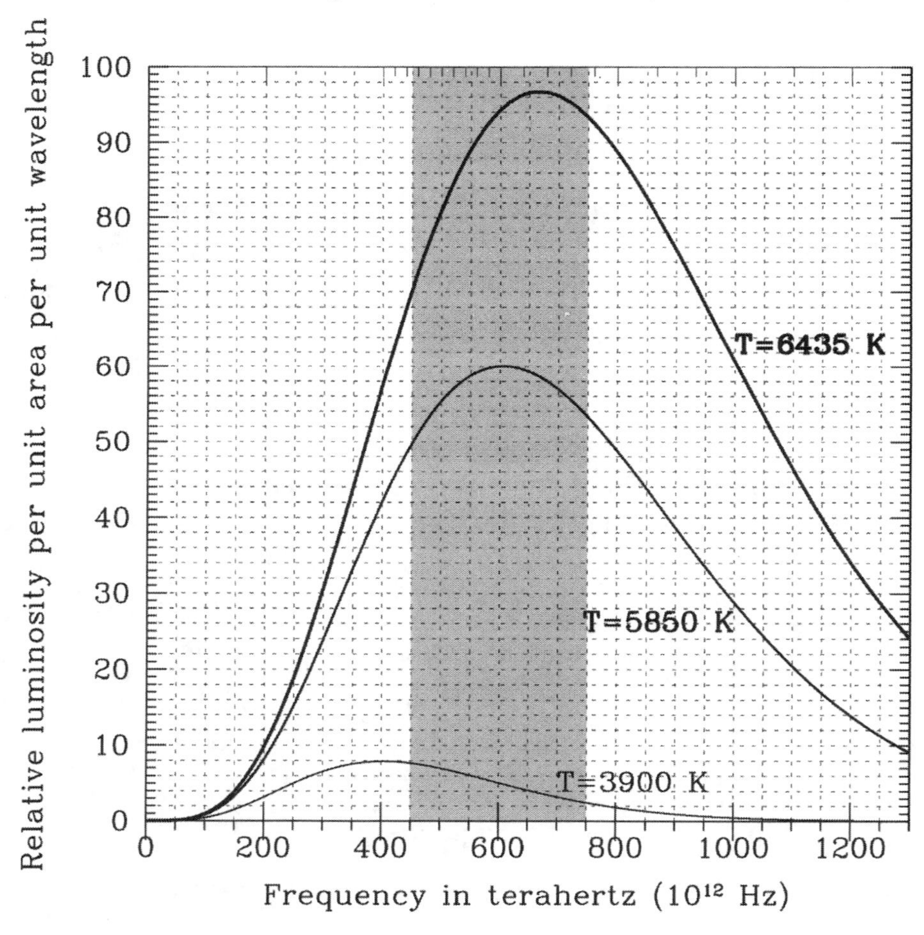

Figure 1: Thermal spectra for three objects, with each object's temperature shown.

Unit 6.2 A Scale Model of The Solar System

Objective

To demonstrate the various sizes and orientations of the Solar System's eight planets as well as lay out the groundwork for constructing a scale model Solar System extending out to Neptune

Introduction

The radius of a planet is a measure of a planet's size, from its uppermost solid layer (or the top of its atmosphere for a gas giant without a discernible solid surface) to the center of its core. Most of the planets are very close to spheres, but the rate at which a planet spins along with its composition causes the planets to bulge slightly at their equators. The most striking example is Saturn, with its very, very low density (about half the density of water) and very fast rotation rate (about 18,000 kilometers per hour) causing the planet's equator to bulge out 6000 kilometers more than its north pole to south pole distance.

Another pronounced characteristic of a planet is its axial tilt. The axial tilt of a planet describes the angle that the planet's equator makes to the planet's orbital path. For a planet with no axial tilt, the line of its north-south pole axis would be perpendicular (90°) to its orbital plane. The equator of an un-tilted planet would lie directly on top of the planet's orbital path. For various reasons – some lost to ancient history; several of the planets are substantially tipped and appear to lean toward or away from the Sun during their orbit, causing their north-south pole line to be bowed toward the path of their orbit. Earth's 23.5° tilt causes the rise of the seasons: when the Earth's orbital position leaves its north pole bowed toward the Sun, we experience the warmer summer months in the northern hemisphere; when Earth's pole is tilted away from the Sun, the northern hemisphere receives less light and experiences the cooler winter months.

Scale models are crucial to the world of science, engineering, architecture, and civil planning. A scale model is a perfectly proportioned miniature model of a much larger object. Before a building is constructed on campus, before the frame of a concept airplane is built, or before a car rolls off the assembly line, a physically smaller but proportionally identical model is created for testing and experimental purposes. The sizes and distances of Solar System objects make their scale difficult to grasp. Given the titanic size and scale of the Solar System, creating a properly scaled model is difficult. Either the planets are large and easily visible and the distances between the model planets are many miles across or the model is compact and small but the planets are microscopic.

However, in the same way as an architect may build a perfectly scaled miniature of a building to demonstrate what the structure will eventually look like, reproducing a scale model of the Solar System will give you insight into both the proportional size of the planets and their properly proportional distances. A scale model takes a size measurement and shrinks it down for all sizes and distances, equally. Unlike an elementary school science fair, where small Styrofoam balls all fit on a small table lined up with one another, you will quickly realize that even a scale model with a small scale can become incredibly large very quickly.

To calculate how large a scaled model must be, there is a need to relate real-world distances with scale model distances. In the case of a scale model, a planet with a diameter many thousands of kilometers wide may be shrunken down to just a few centimeters.

Equations and Constants

Equation	Expression	Variables
Scale Factor	$Scale\ Factor = \dfrac{Real}{Ruler}$	*Scale Factor*: a conversion factor which bridges actual and scale model sizes *Real*: a real-world distance or size,such as kilometers or AU *Ruler*: a distance or size used in a scale model,such as meters or centimeters

Illustrations

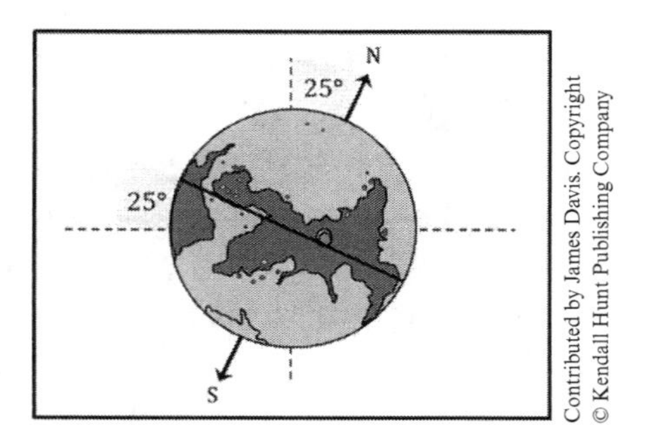

Figure 1. A schematic of the axial tilt of Mars. The dark black line represents Mars's equator with the N-S line representing the planet's polar axis, about which the planet rotates. The dashed horizontal line is the orbit of Mars, the route that it travels around the Sun, and it points directly at the Sun. The dotted vertical line is perpendicular to Mars's orbit and represents the location of Mars's north-south pole line if the planet had zero degrees of axial tilt. Instead, Mars is tilted 25° so its equator rises 25° above its orbital path and its axis likewise leans 25° away from the perpendicular line. The north-south spin axis and the equator are always perpendicular (i.e., they make a 90° angle to one another).

Procedure

The first part of the lab involves drawing accurate representations of the eight planets, including their proper scaled diameters and axial tilts (i.e., their tilt measured with respect to the planetary orbital plane). Earth is 6,370 kilometers in radius, the distance from the core to the surface, with a diameter or 12,740 kilometers. In this model, 6,370 kilometers is set equal to 1.00 centimeter on paper, and all of the planets are drawn relative to this smaller scale. Your ScaleFactor, therefore, would be 6,370 km/cm. You may choose to use a different scale, however.

Using the Scale Factor equation and the actual radii of the eight planets and the Sun, calculate the scale model radius of the given objects, in centimeters. Draw these planets on Worksheets #3 through #7 using the steps listed below. On Worksheet #2, you will properly scale the distance between the miniature planets. Using the same scale factor as on Worksheet #1, convert the Sun-to-planet distances from kilometers into centimeters, using the same process as done on Worksheet #1. Since the numbers will be very large, convert the distances into meters. Using a map, computer program, or a list of locations and distances, choose a location for the center of the Solar System (the Sun). With the scale model distances, build a scale model of the Solar System by determining where the planets should be located (by naming a landmark or intersection that given distance from the Sun). To properly draw the size and orientation of the planets, follow the steps below and refer to the figures:

Step 1: On Worksheets #3 through #7, the orbital plane of each planet has been drawn and labeled. Place a very small **x** mark or dot on the dashed Orbit Line to mark the exact physical center of the planet. Using a ruler, draw a line from the **x** along the orbit with a length equal to the planet's scaled radius in centimeters from Worksheet #1.

Step 2: Using your compass, place the pencil end on the end of your radius line and the metal compass tip on the marked planetary center. Draw a smooth circle representing the planet's circumference. It is usually easier to actually hold the compass still and turn the paper, rather than trying to spin the compass itself around. The finished circle represents the full disk of each planet.

Step 3: Each of the planets has a distinct axial tilt. Using your protractor, place the center mark on the marked planetary center and place a small dot at the angle measurement corresponding to the planet's tilt. Draw a line across the planet's midsection. This is the geometrical equator of the planet and has a length equal to the scaled planet's diameter.

Step 4: Now line the protractor up on the equator – laid on the center mark again – and mark a small point at 90°. Draw a second straight line completely through the planet, perpendicular to the axis. This represents the north-south axis of the planet. Label the north pole with an N and the south polewith an S.

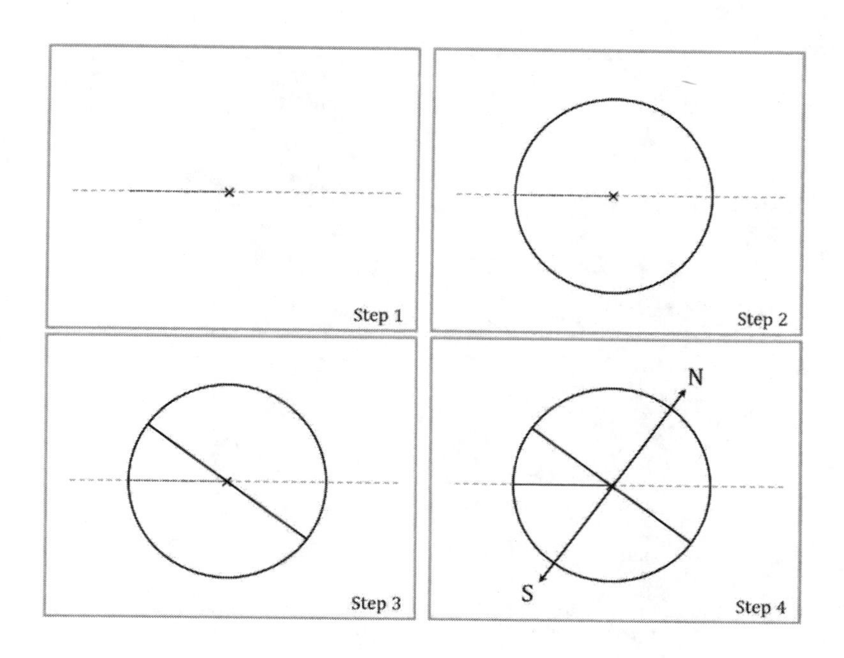

NAME ______________________ ID ______________________

DUE DATE __________ LAB INSTRUCTOR __________ SECTION __________

Worksheet # 1

Scale Factor Used (in km/cm)

Star	Radius (in km)	Scaled Radius (in cm)	Axial Tilt (in degrees)
Sun	695,500		0°

Planet	Radius (in km)	Scaled Radius (in cm)	Axial Tilt (in degrees)
Mercury	2,440		0°
Venus	6,050		177°
Earth	6,370		23.5°
Mars	3,390		25°
Jupiter	69,900		3°
Saturn	58,200		27°
Uranus	25,400		98°
Neptune	24,600		29°

Note: The axial tilt measures the angle between the plane of a planet's orbital plane and its equator.

NAME ____________________ ID ____________________

DUE DATE __________ LAB INSTRUCTOR __________ SECTION __________

Worksheet # 2

Planet	Distance (in km)	Distance (in AU)	Scaled Distance (in m)	Placement in Model
Mercury	57,910,000	0.387		
Venus	108,208,000	0.723		
Earth	149,598,000	1.00		
Mars	227,939,000	1.52		
Jupiter	778,547,000	5.20		
Saturn	1,433,449,000	9.58		
Uranus	2,870,671,000	19.2		
Neptune	4,498,542,000	30.1		

NAME ______________________________ ID ______________________________

DUE DATE __________ LAB INSTRUCTOR ___________ SECTION __________

Worksheet # 3

The Inferior Planets: Mercury and Venus

Orbit

Line

Orbit

Line

NAME ______________________ ID ______________________

DUE DATE __________ LAB INSTRUCTOR __________ SECTION __________

Worksheet # 4

The Outer Terrestrial Planets: Earth and Mars

Orbit
Line

Orbit
Line

NAME ______________________ ID ______________________

DUE DATE __________ LAB INSTRUCTOR __________ SECTION __________

Worksheet # 5

The Gas Giant: Jupiter

Orbit
Line

NAME ______________________ ID ______________________

DUE DATE __________ LAB INSTRUCTOR __________ SECTION __________

Worksheet # 6

The Gas Giant: Saturn

Orbit

Line

NAME ______________________________ ID ______________________________

DUE DATE __________ LAB INSTRUCTOR __________ SECTION __________

Worksheet # 7

The Ice Giants: Uranus and Neptune

Orbit
Line

Orbit
Line

NAME ______________________ ID ______________________

DUE DATE __________ LAB INSTRUCTOR __________ SECTION __________

Worksheet # 8

Postlab Questions

For each of the following questions, include all work, equations, and proper units.

1. The severity of seasonal temperature swings – summer to winter – is based on the tilt of a planet's axis toward (in summer) and away from (in winter) the Sun. The larger the tilt, i.e., the closer the poles lie to the orbital plane, the more extreme seasons are expected. Which *single planet* in the Solar System would have the most severe seasons in this case? Which *planets* would have the least severe seasons?

2. Saturn's brightest sets of rings, from one edge to the other, are 280,000 km wide. How many centimeters across would you draw the rings in your scale model?

3. The closest star to the Earth besides the Sun is the red dwarf star Proxima Centauri, located at a distance of 4.26 light years (269,000 AU). Using the same scale as used for the Solar System, how far away should Proxima Centauri be placed in a scale model (include units).

4. If 1 mile = 1610 meters, how far away should Proxima Centauri be from the scale model Sun (include units)?

5. The furthest apart one can place two objects on planet Earth is about 8,000 miles (two opposite sides of the planet). What is the immediate problem you see with building a scale model encompassing the Earth, planets, and Proxima Centauri?

Unit 6.5 Kuiper Belt Objects

Objective

To learn about the properties of the Kuiper Belt Objects orbiting beyond Neptune, including their physical sizes, orbital eccentricities, and orbital inclinations

Introduction

Following the discovery of Uranus in 1781 and Neptune in 1846, astronomers searched for signs of other planets lurking in far reaches of the Solar System on long-period orbits for centuries. It has not been until Clyde Tombaugh, working at the Lowell Observatory (Flagstaff, Arizona), spotted a small pinpoint of moving light on a photographic plate in 1930. A new planet exterior to the orbit of Neptune had been discovered. The astronomer Gerard Kuiper theorized that the formation of the Solar System would leave a ring of icy, leftover material in the far reaches of the Solar System, from just beyond Neptune's orbit (~30 AU) out to about 55 AU. Like the asteroid belt between Mars and Jupiter, the Kuiper Belt would be a wide disk of planetary debris slowly orbiting the Sun. It was from this location (aside from the Oort Cloud) that the Solar System's comets would originate: small chunks of ice and rock that vaporize spectacularly as they approach the Sun. Kuiper theorized that many billions of tiny icy fragments may orbit out in that vast region, with a comet created every time a collision or gravitational nudge disturbed the orbit and caused it to careen toward the Sun.

The first Kuiper Belt Object (KBO) fitting this description was 1992 QB_1, discovered in 1992, showing for the first time that there were objects in the outskirt of the Solar System, inhabiting the same region of space as Pluto. In 2005, the object 2003 UB_{313} was identified to be about the same size as but more massive than Pluto, eventually dubbed Eris. Eris joined other substantial KBOs, like Makemake and Haumea. Taken together, these objects showed that Pluto composed only a tiny fraction of the Kuiper Belt's mass, representing a less dominant object than any of the Solar System's eight planets and more a primitive, unfinished remnant of planetary formation. Pluto was subsequently reclassified from planet to dwarf planet.

Today, over 1000 KBOs have been confirmed, with enough observations to allow astronomers to plot out and predict their orbits. The simple Kuiper Belt turned out to be more complex than originally thought, consisting of different, distinct families of KBOs. The objects called Plutinos (like, nor surprisingly, Pluto) show effects of being strongly influenced by Neptune's gravity. Plutinos have perihelion distances which are very close to Neptune's semi-major axis distance of 30.1 AU. In addition, Plutino orbits synchronize with Neptune's orbit, with lengths that are 1.50, 2.00, or 2.50 times the length of Neptune's 164.8 year orbit (i.e., orbits of 248 years, 330 years, or 412 years, respectively).

A second class of KBO, called Cubewanos (like the namesake object QB_1) have orbits that are unperturbed and unaffected by Neptune. Their eccentricities are very small, usually less than 0.1, and have perihelion distances that are nowhere near Neptune's 30.1 AU semi-major axis. Finally, the Kuiper Belt is populated by some objects showing a history of having been badly disturbed into elongated, chaotic orbits by interactions with Neptune or some other distant object. These bodies are called Scattered Disk Objects (SDOs). Scattered disk KBOs are marked by high inclinations, with tilted orbits that lie high above or below the orbits of the eight planets, along with aphelion and perihelion distances that vary greatly and never get very close to Neptune's orbit (with eccentricities larger than 0.25). The scattered disk serves as evidence that the early Solar System was a more disturbed, chaotic place, with these KBOs serving as reminders of the past upheaval.

Equations and Constants

Equation	Expression	Variables
Aphelion	$Q = a(1+e)$	Q: the aphelion distance a: the semi-major axis of the orbit e: the eccentricity of the orbit
Perihelion	$q = a(1-e)$	q: the perihelion distance a: the semi-major axis of the orbit e: the eccentricity of the orbit
Kepler's Third Law	$a^3 = P^2$	a: the semi-major axis of the orbit in AU P: the period of the orbit in years

Procedure

For Worksheet #1, you will identify the largest KBOs by their size. Refer to Datasheet #1 as well as Table 3 in the Appendix for information. The largest circle (marked by a dotted line) represents the circumference of Mercury, the Solar System's smallest planet. The other circles (labeled 1 through 11) represent the 11 largest known KBOs (including Pluto's largest moon, Charon). Despite its small size, Mercury's diameter of 4880 kilometers is over twice the diameter of the largest KBO, Pluto. The drawings of the KBOs' circumferences are scaled so that 1.0 millimeter in the drawing is equal to 36 kilometers of actual length. Because many KBOs have remarkably similar diameters, which appear almost identical in the drawings, several of the objects have already been identified on Worksheet #1. By measurement and elimination, determine which circles correspond to the physical sizes of the remaining 7 KBOs.

Also on Worksheet #1, choose 5 KBOs (excluding Sedna and Pluto's moon Charon). Enter their names. From the Appendix, enter the semi-major axis of the KBO and its eccentricity and calculate both its aphelion distance (in AU) and its perihelion distance (in AU).

Part of Pluto's reclassification from planet to dwarf planet was based on its inability to "clear its neighborhood" of other KBOs. This means that Pluto's orbital region overlaps with the orbital regions of numerous other KBOs. On Worksheet #2 there are five columns, each with a space below to record the name of a KBO from Worksheet #1 and with an axis labeled with distances running from 0 AU (the location of the Sun) to 100 AU (far beyond the outer edge of the Kuiper Belt). The shaded horizontal boxes represent the aphelion and perihelion distances of the eight Solar System's planets, showing the domain dominated by them. Note that Uranus is the most eccentric planet, with a widely spaced perihelion and aphelion distance. Also note that none of the eight planets' domains overlap (or even draw close to one another). Using your KBOs and their calculated aphelion and perihelion distances from Worksheet #1, plot a pair of lines across the single column to represent the perihelion and aphelion distance of each KBO. Shade in the space between those lines. This represents the "neighborhood" of the KBO. For a planet, that neighborhood would be free of any overlapping neighborhoods from the other columns. This will clearly not be the case for the KBOs, as you may notice significant regions of overlap.

For Worksheet #3, use your five chosen KBOs again and look up their orbital inclinations (the tilt of their orbit relative to the plane of the Earth's orbit around the Sun). Mark with a small dot the angular position of those KBO's orbits, connect the dot to the location of the Sun with a dashed line (representing the plane of the planet's orbit) and label the point with the KBO's name.

Datasheet #1

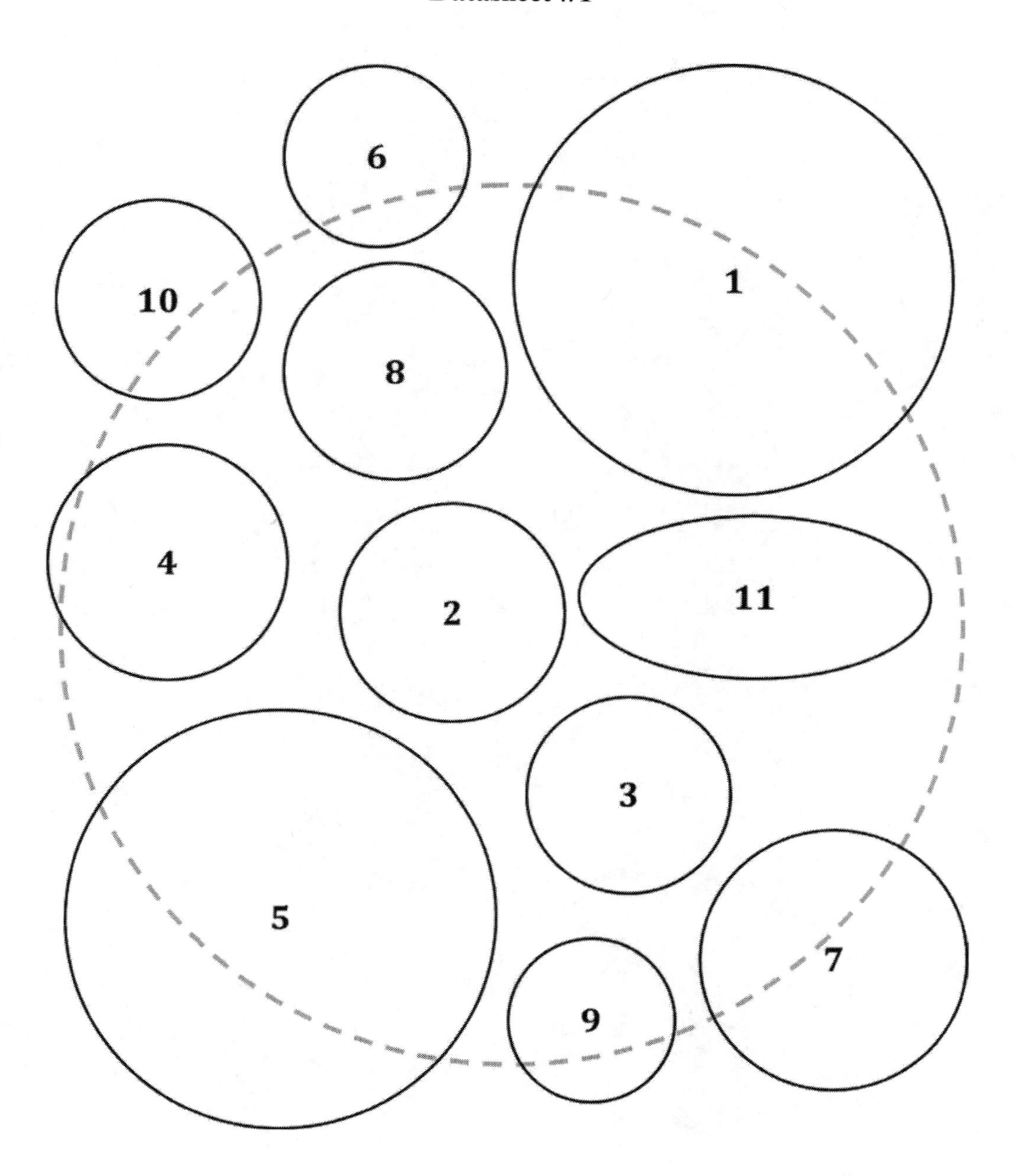

NAME ______________________ ID ______________________

DUE DATE __________ LAB INSTRUCTOR ______________ SECTION __________

Worksheet # 1

KBO Sizes

Object	KBO Name	Measured Diameter (in mm)	Calculated Diameter (in km)
1	Pluto	66	2370
2	Charon	34	1200
3	Orcus	31	1110
4	2007 OR_{10}	36	1290
5			
6			
7			
8			
9			
10			
11			

KBO Orbits

KBO Name	Semi-Major Axis (in AU)	Eccentricity	Perihelion (in AU)	Aphelion (in AU)

Data obtained from:

http://www2.ess.ucla.edu/~jewitt/kb/big_kbo.html

NAME ______________________________ ID ______________________________

DUE DATE __________ LAB INSTRUCTOR ______________ SECTION __________

Worksheet # 2

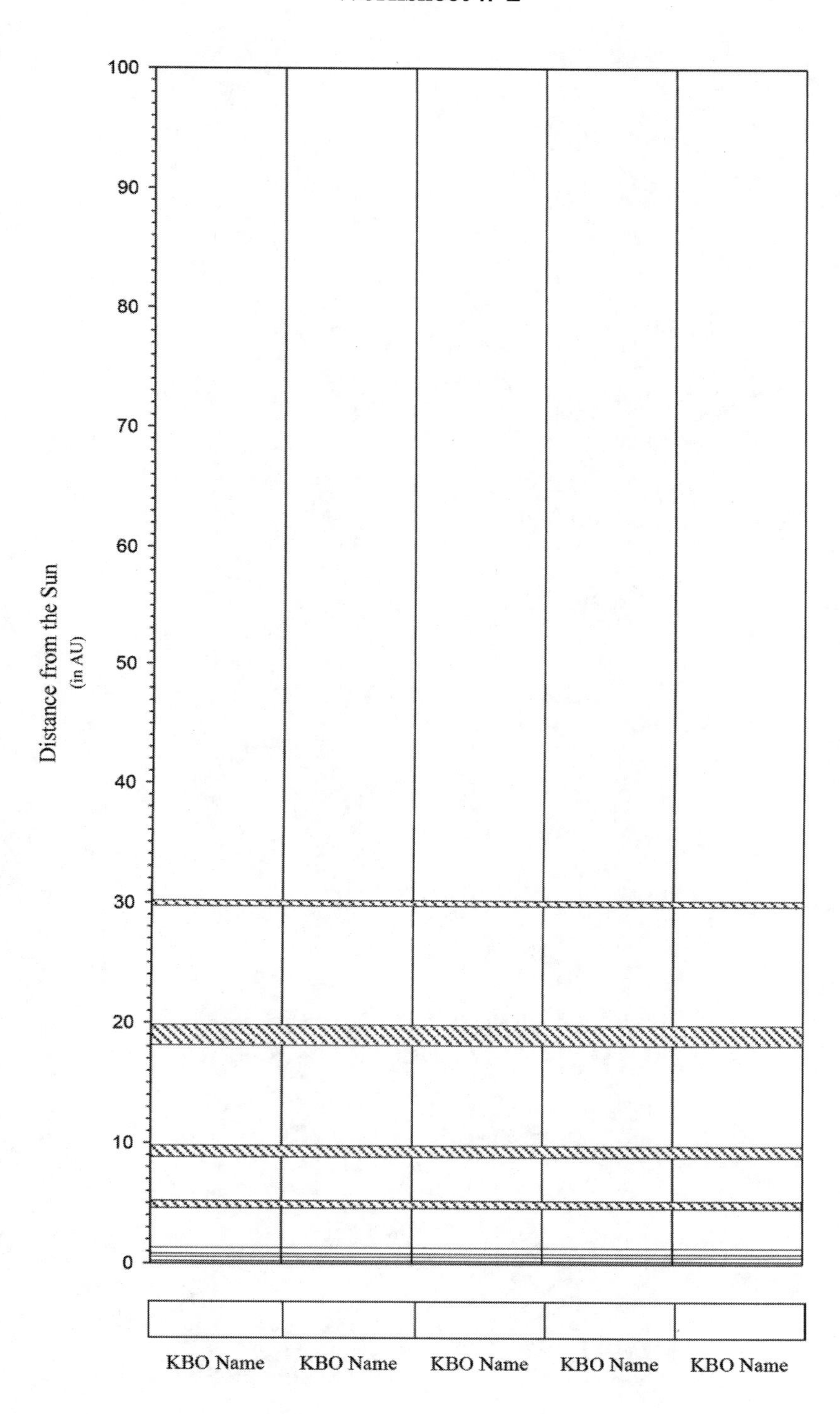

NAME ______________________________ ID ______________________________

DUE DATE __________ LAB INSTRUCTOR ______________ SECTION __________

Worksheet # 3

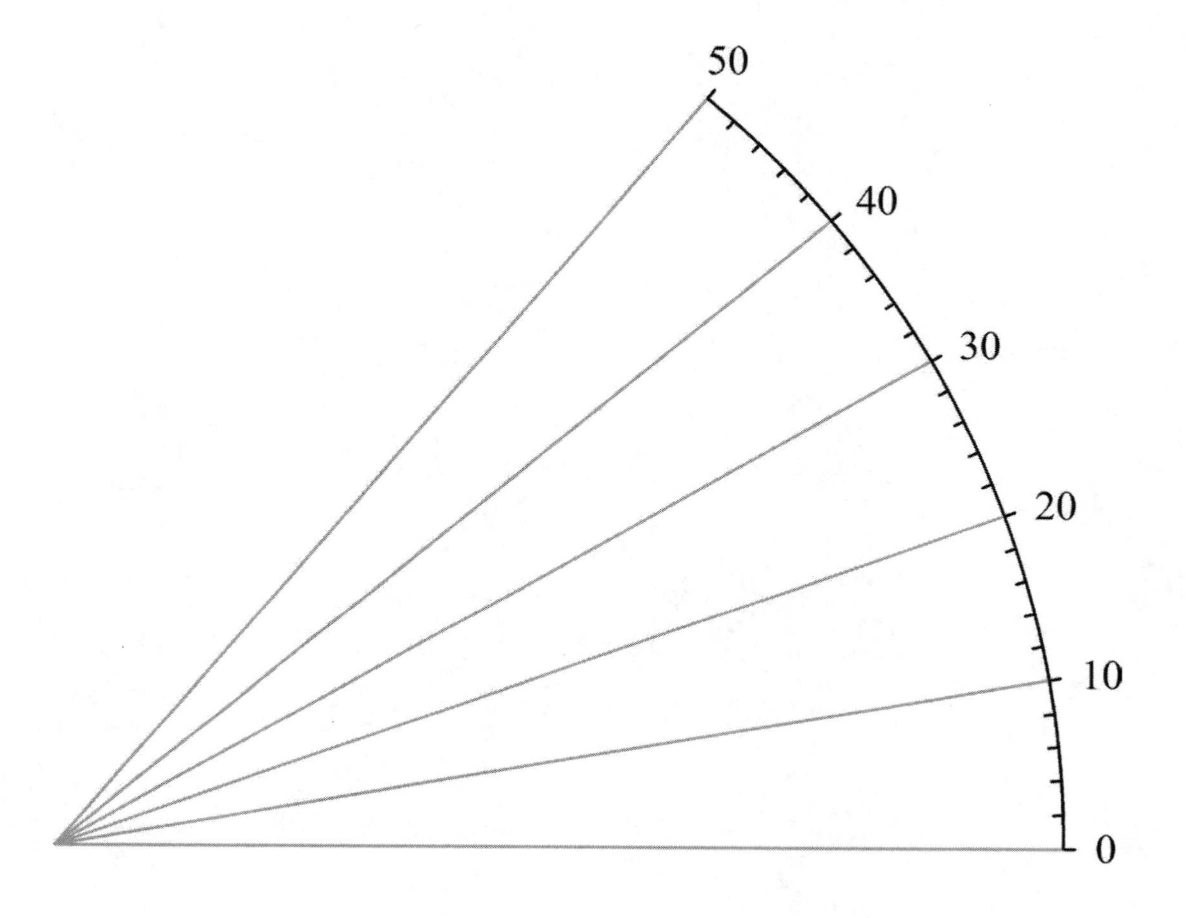

NAME ______________________________ ID ______________________________

DUE DATE __________ LAB INSTRUCTOR ______________ SECTION __________

Worksheet # 4

KBOs fall into three classes, as covered in the Introduction: cubewanos, plutinos, and SDOs. Using the data for the KBOs from the Appendix Table 3, answer the following questions about the nature of some KBOs. Calculate the necessary values of perihelion, period, and period-ratio or state pertinent values given in the KBO data table, such as inclination or eccentricity. Points are awarded on thoroughness of answers, so be as complete as possible in your replies.

1. Referring back to the orbital inclinations of your KBOs as selected, why even if two KBOs have overlapping "neighborhoods" or share an orbit, they will be very unlikely to crash into one another?

2. What family of KBOs does Orcus belong to? Is it a Cubewano, a Plutino, or an SDO? Show calculations or measurements that lead to this conclusion and explain your answer.

3. What family of KBOs does Quaoar belong to? Is it a Cubewano, a Plutino, or an SDO? Show calculations or measurements that lead to this conclusion and explain your answer.

4. What is Eris's orbital period, in years?

5. Today, Sedna is about as close to the Sun as its orbit allows it to be, which allowed astronomers to detect it before its long orbit takes it far out of the Kuiper Belt and makes it invisible to most telescopes. What is Sedna's period, in years?

6. In 2014, astronomers discovered 22 new KBOs, among those the following three: 2014 UF_{224}, with a period of 331.1 years; 2014 TT_{85}, with a period of 280.8 years; 2014 QM_{441}, with a period of 315.2 years. Given that information and knowing that Neptune's period is 164.8 years, which of those objects – if any – are likely part of the Plutino family of KBOs? Show your work that lead you to your conclusion.

Age Dating Through Radioactive Decay

NAME ______________________ ID# ______________________

DATE ______________________

An atomic nucleus consists of particles known as protons and neutrons. Every nucleus of a given **element** has the same number of protons: hydrogen has one proton, helium has two protons, etc. Nuclei of the same element with different numbers of neutrons are known as **isotopes**. For example, deuterium (1 proton + 1 neutron) and tritium (1 proton + 2 neutrons) are isotopes of hydrogen.

If an atomic nucleus has too few or too many neutrons relative to its number of protons, it will be **unstable** and will decay into a different nucleus; this is called **radioactive decay**. A nucleus with too many neutrons can decay when one neutron turns into a proton and an electron. For example, tritium decays into helium-3 (2 protons and 1 neutron) and an electron.

Each unstable isotope has a 50% chance of decaying within a time span known as that isotope's **half-life**.

A single unstable isotope can decay at any time, but in a large sample of nuclei of that isotope, you will find that 50% of the nuclei will decay during the first half-life you study, 50% of the remaining nuclei will decay during the next half-life you study, 50% of the remaining nuclei will decay during the next half-life after that, etc.

Consider a specific (but fictonal) example. Element Tr (for Triangle) decays to element Fc (filled circle). In Figure 1, 16 newly created nuclei of element Tr are shown at time zero on the left. (The nuclei might be newly created in a supernova, for example.) Each Tr nucleus is shown as a triangle inside a circle. In the same Figure, those same 16 nuclei are also shown at four **equally spaced** later times. Use Figure 1 to answer the following questions.

1. At time 1, how many nuclei of element Tr are left? ____________
2. What fraction of the original nuclei of element Tr are left at time 1? ____________
3. What do we call the amount of time between time 0 and time 1? ____________
4. Complete Figure 1 by drawing in the correct number of Tr nuclei (circled triangles) and Fc nuclei (filled circles) at times 2, 3 and 4.
5. Fill in the table below with the number of nuclei of element Tr and of element Fc at each time. Then find the ratio of the number of nuclei of element Fc to the number of nuclei of element Tr. That ratio is what is used to estimate the age of a sample of material containing an element and the radioactive decay product of that element. Fill in that column of the table.

Time	# of Tr nuclei	# of Fc nuclei	Ratio Fc/Tr (Product Isotope/Parent Isotope)
0	16	0	0
1			
2			
3			
4			

6. You observe a certain star and find that by number, 0.1% of its atmosphere is made of isotope-X and 0.7% is made of isotope-Y. Isotope-X is an unstable isotope that decays to isotope-Y with a half-life of 1.5 billion years, and that is the only way to produce isotope-Y. How old is the star? ______________________________

The next two questions are about carbon dating. Trees continuously absorb the unstable isotope carbon-14 from the atmosphere. When a tree is chopped down, no more carbon-14 is absorbed, and the existing carbon-14 decays to nitrogen-14 with a half-life of about 6000 years.

7. A sample of a piece of burnt wood from a cave in Freedonia contains 14 to 15 parts per million of nitrogen-14 and 1 to 2 parts per million of carbon-14. What is the range of ages possible for when the tree was chopped down?

____

8. A new, more sensitive analysis is applied to another small sample of the same piece of wood. It is found to contain 15,000 to 15,015 parts per billion of nitrogen-14 and 1,000 to 1,001 parts per billion of carbon-14. What is the revised range of ages possible for this charcoal?

____

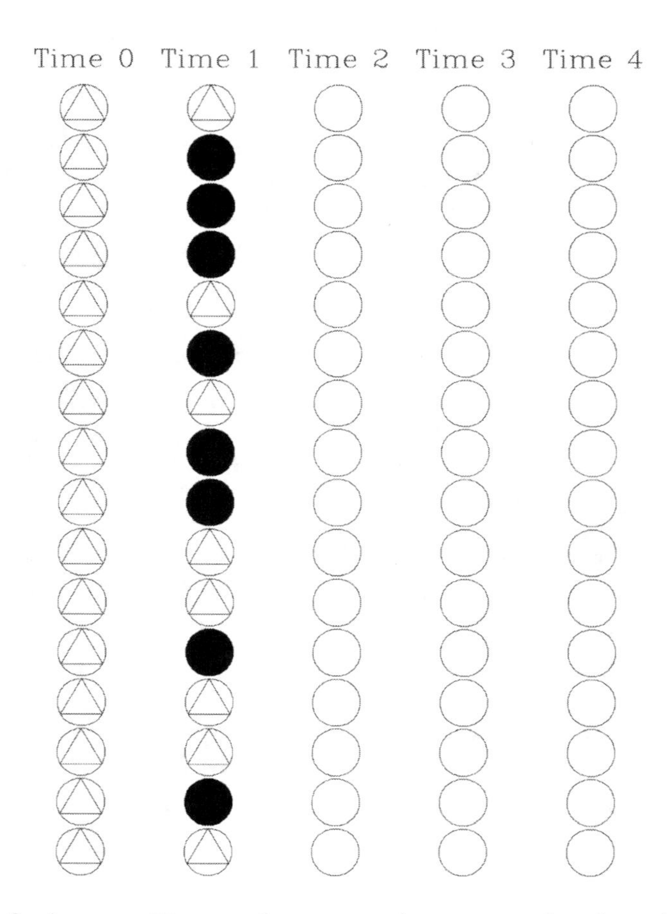

Figure 1: Sixteen nuclei of element Tr are shown at time zero. At time 1, some nuclei have radioactively decayed to element Fc (filled circles). The situations at later times are left blank.

Tidal Forces and the Roche Limit

NAME ______________________ ID# ______________________

DATE ______________________

Imagine a solar system with a planet orbited by two moons (Figure 1). The moons are identical except that one orbits farther from the planet than the other. The center of moon 1 is located 14 moon radii from the center of the planet, and the center of moon 2 is located 9 moon radii from it. Each moon has mass m=20, the planet has mass M=10,000, and the force of the planet's gravity *on an object of mass m=1* (for Newton's G=1) has been calculated at each distance in Figure 1. For example, the force of gravity on an object of mass m=1 at the center of moon 1, a distance d=14 from the center of the planet, is $F = \frac{G \times M \times m}{d \times d} = \frac{1 \times 10{,}000 \times 1}{14 \times 14} = 51$.

1. The orbital motion of each moon balances the force of the planet's gravity as measured at the *center* of each moon, which is why the moons stay in orbit instead of falling toward the planet. However, this orbital motion does not exactly balance the force of the planet's gravity at the *surface* of each moon. Those unbalanced forces are called **tidal forces**, and we can use arrows to see their effects. *Fill in the blank lines next to each moon in Figure 1 by summing the two numbers to the left of each line. The resulting numbers are the tidal forces at points A and B on moon 1 and at points C and D on moon 2*. The calculations at the center of each moon have been done for you: both moon 1 and moon 2 experience zero tidal force at their centers.

2. Let's compare those tidal forces to the force of gravity at the surface of each moon. In question 1 you calculated tidal forces at points A, B, C, and D. *Now draw arrows of those lengths at A, B, C, and D, pointing in the appropriate direction for a negative or positive force*. **Useful information:** • Each moon's force of gravity (on an object of mass m=1) is shown by the arrows already drawn at points A, B, C, and D, pointing toward the center of each moon. • The length of each arrow represents 20 force units. • When a force points toward the planet, it is negative. • When a force points away from the planet, it is positive.

3. In what direction relative to the *surface* of moon 1 is the tidal force trying to move objects at point A? ____________ At point B? ____________

4. In what direction relative to the *surface* of moon 2 is the tidal force trying to move objects at point C? ____________ At point D? ____________

5. In Figure 1, a longer arrow means a stronger force. Is the tidal force stronger than the moon's gravity at point A on moon 1? __________ ...at point B on moon 1? __________ ...at point C on moon 2? __________ ...at point D on moon 2? __________

6. At any point where you answered no to question 5, the moon's surface will just be pulled a bit in the direction of the tidal force. *If you answered no for moon 1 or moon 2, draw the resulting shape of the surface of the moon(s)* under "Tidally distorted moons and planet" in Figure 1. The dashed circles show the original shapes of each object, and the sides of each tidally distorted object have been drawn in for you as short solid lines. Which is most

distorted by tidal forces: moon 1, moon 2, or the planet? __________________ Which is least distorted? ______________________

7. At any point where you answered yes to question 5, objects on the surface of the moon will be pulled off into space by the tidal force. If the tidal force is much stronger than gravity, then the surface of the moon itself will be pulled off, and the moon will eventually be pulled apart. *Mark any point on moon 1 or moon 2 where the tidal force is larger than the moon's gravity.*
8. We define the **Roche limit** as the orbital radius at which the tidal force on both sides of an object is greater than the object's gravity. At what distance from the planet is the Roche limit for these moons located? __________________ *Draw it in on Figure 1!*

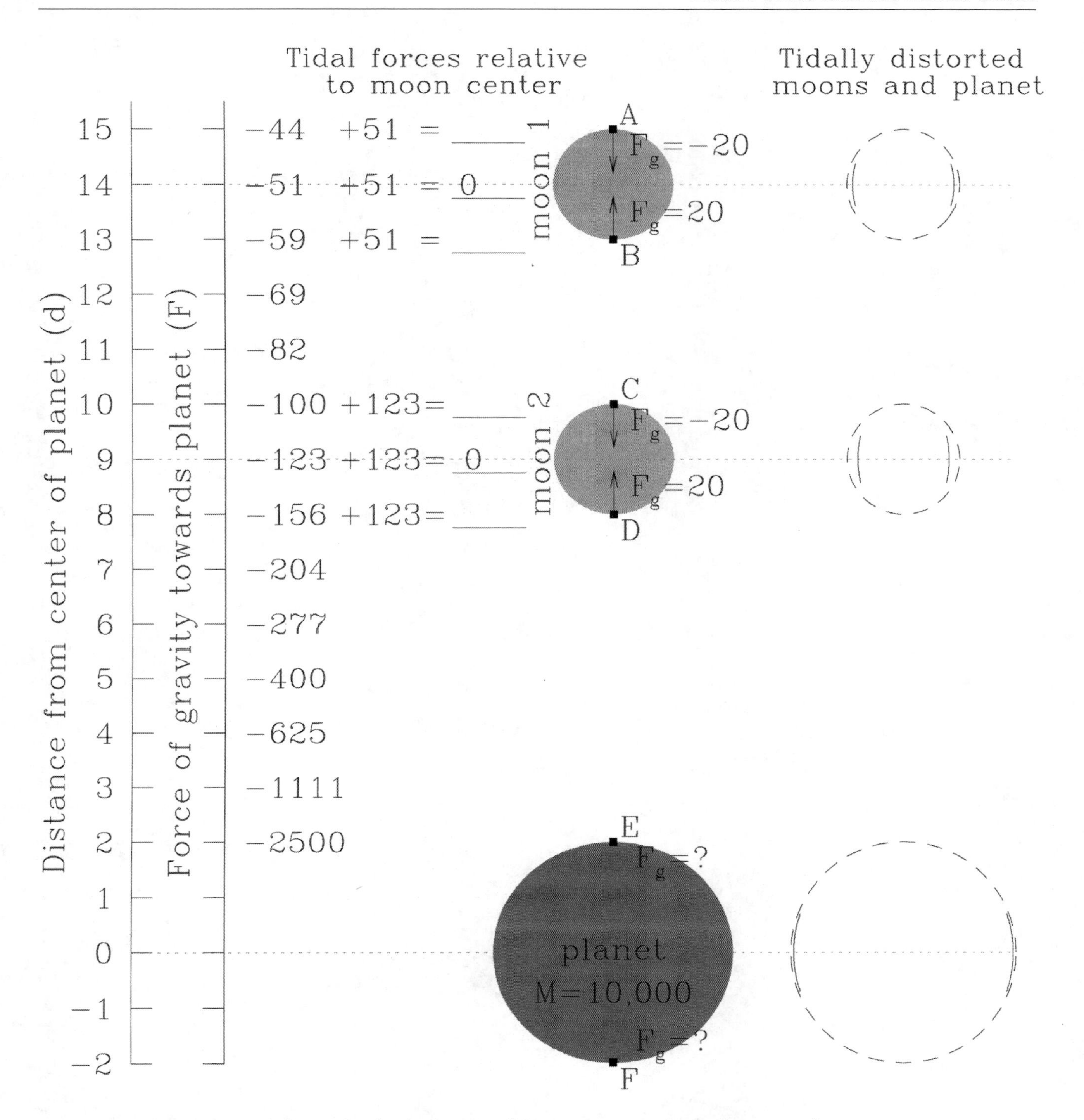

Figure 1: A hypothetical planet with two moons in a distant solar system.

Stellar Masses with Newton's Version of Kepler's Third Law

NAME ______________________________ ID# ______________________________

DATE ______________________________

The centre of mass of two orbiting objects always lies on a line drawn between those objects, and never moves. In a **binary star** system, two stars make elliptical orbits with the centre of mass at one focus. *If two objects have masses* M_1 and M_2 *measured in solar masses* (our Sun has 1 solar mass) and have orbits of semi-major axis a and period *P* around the centre of mass, then Newton's version of Kepler's Third Law is:

$$(\text{Orbital period in Earth years})^2 = \frac{(\text{semi-major axis in AU})^3}{M_1 + M_2} \longrightarrow \boxed{P^2 = \frac{a^3}{M_1 + M_2}} \qquad (1)$$

Also, the distances d_1 & d_2 of objects 1 & 2 from their centre of mass are related by $\boxed{\frac{d_1}{d_2} = \frac{m_2}{m_1}}$ where *the distance between the objects is* $\mathbf{d = d_1 + d_2}$. (If the objects were weights connected by a stick, the centre of mass is where you could balance the stick on your finger.)

Let's draw centres of mass and orbits in Figure 1 using Newton's version of Kepler's 3rd law. This has been done for you in the leftmost system; read the figure caption for the details. And in systems I and II, *the dashed line shows the orbit the light grey object would have if the dark grey object had 1000 times the mass of the light grey object.*

1. In system I, objects A and B have the same mass.

   Mark the centre of mass. Assuming circular orbits, draw the orbits of both objects.

2. In system II, object C has three times the mass of object D.

   Mark the centre of mass. Assuming circular orbits, draw the orbits of both objects.

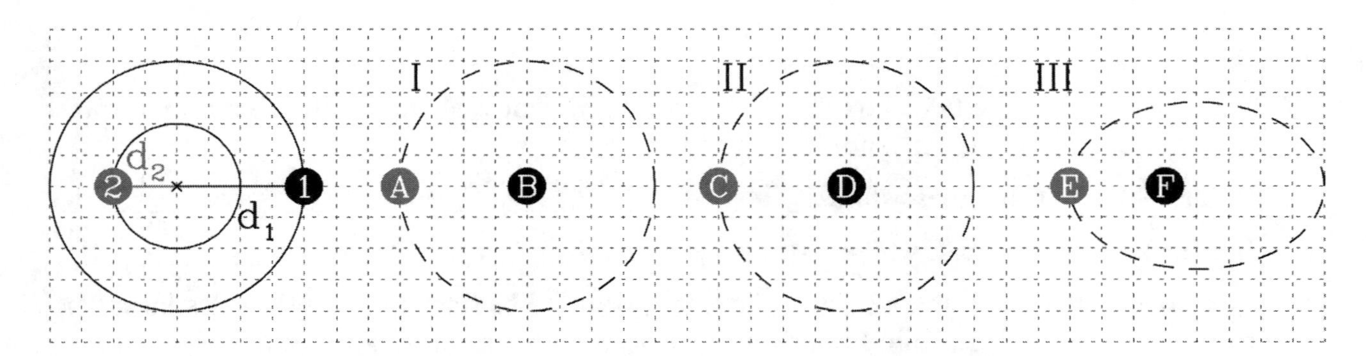

Figure 1: Objects orbiting their centres of mass. In the leftmost system, object 2 has twice the mass of object 1, so its distance from the centre of mass (the ×) is $d_2 = \frac{1}{2}d_1$. *It is always true that the more massive object traces out a smaller orbit than the less massive object does.*

3. In system III, objects E and F have the same mass. The elliptical orbit of object E around the centre of mass is shown. Mark the centre of mass and draw in the orbit of F.

Finding the Masses of Orbiting Objects

Newton's version of Kepler's Third Law enables us to determine the mass of **any** objects orbiting each other (stars, for example) **if** we can determine their orbital period and semi-major axis. *If two objects have masses M_1 and M_2 measured in solar masses* (our Sun has 1 solar mass) and have orbits of orbital period P and semi-major axis a around the centre of mass, then Newton's version of Kepler's Third Law is:

$$(\text{Orbital period in Earth years})^2 = \frac{(\text{semi-major axis in AU})^3}{M_1 + M_2 (\text{in solar masses})} \longrightarrow \boxed{P^2 = \frac{a^3}{M_1 + M_2}} \qquad (2)$$

For elliptical orbits, the separation d changes with time but a is always given by : $a = \frac{d_{min} + d_{max}}{2}$ where d_{min} and d_{max} are the minimum and maximum separation of the objects.

4. On the previous page, if stars A and B in system I take 2 years to orbit each other, how much mass do each of them have?

____

5. On the previous page, if stars C and D in system II take 1 year to orbit each other, how much mass do each of them have?

____

6. On the previous page, if stars E and F in system III take 2 years to orbit each other, how much mass do each of them have?

____

7. Suppose you observe *two identical stars, each with half the mass of our Sun,* which orbit their center of mass with a semi-major axis of 1 AU. How many years does it take them to complete one orbit?

____

8. Suppose you observe two stars (1 and 2) orbit their centre of mass with a semi-major axis of 2 AU, taking 4 years to complete one orbit.

   What is $M_1 + M_2$ for these stars? ______________________

   If the stars are identical, what is M_1 and what is M_2? ______________________

9. You observe two stars and find that one star appears to be stationary on the celestial sphere while the other moves around it in a circular orbit with a semi-major axis of 4 AU, making one orbit every 4 years. What is the mass of the apparently stationary star?

____